ARE WE FREAKS OF NATURE?

A New View of Evolution

ARE WE FREAKS OF NATURE?

A New View of Evolution

by

BRANKO BOKUN

TIGER & TYGER
London
MM

PUBLISHED IN THE UK BY

TIGER & TYGER

7 Lower Grosvenor Place
London SW1

DISTRIBUTED BY

THOMAS LYSTER LTD

Unit 9 ~ Ormskirk Industrial Park
Old Boundary Way
Burscough Road
Ormskirk
Lancashire L39 2YW
Tel: 01695 575112
Fax: 01695 570120

First Published 2000
© Branko Bokun 2000
ISBN 1 902914 00 7

BY THE SAME AUTHOR

The Pornocracy	Tom Stacey	1971
Spy In The Vatican	Tom Stacey	1973
Man, The Fallen Ape	Doubleday N.Y.	1977
Humour Therapy	Vita Books	1986
Aids, A Different View	Vita Books	1987
Stress Addiction	Vita Books	1989
Self Help With Stress	Vita Books	1991
Matriarchy in Post Capitalism	Vita Books	1994
Humour and Pathos in Judaeo-Christianity	Avon Books	1997

To the memory of Anne Tichborne

Contents

The Origin of the Human Species

For years I have pondered on what an ape's account of the history of mankind might be like. An animal is necessarily closer to nature — its laws and its logic — than a human. Only a being whose innate natural logic had not been flawed by the fanciful elaborations of a sophisticated mind could explain the history of mankind, and explain it in a way that was devoid of wishful beliefs and free from the constraints of the infantile mentality.

As the human mind developed, human beings lost their *natural* logic. Consequently their explanation of natural phenomena, particularly those surrounding their own origin, cannot be anything but biased.

Our ape might begin his history of mankind with modern man, in whose latest incarnation are embodied all the stages of the evolution process. It is unnecessary to speculate about the meaning of fossils. All fossils of the human species can be found in any living human being. Human abstract thoughts are the result of the brain, a brain programmed by the major events in the life of the human species. However proud of our mind we may be, the mind cannot create *ex nihilo*. By analysing the development of human abstract thought we can uncover the major happenings in the evolution of the human species, the happenings which have conditioned or inspired these abstractions.

The study of prehistory, working backwards from modern man, can also be dangerous. Scientists tend to uncover only what suits their beliefs, prejudices, metaphysical preconceptions, or simply their conceit.

Mankind tends to glorify itself and its ancestors. On reading some of the books that have been written on the origin of mankind, one forms the impression that our ancestors grew apart from the apes and faced the vicissitudes of the savannah merely to please the ego of modern man and give him the illusion of superiority.

Konrad Lorenz, in his book *On Aggression*, stressed the following apologetic words: "...and who has gained insight into evolution, will be able to apprehend the unique position of man. We are the highest achievement reached so far by the greatest construction of evolution." Seven pages later Lorenz writes: "Unreasoning and unreasonable human-nature causes two nations to compete, though no economic necessity compels them to do so; it induces two political parties or religions with amazing similar programmes of salvation, to fight each other bitterly and impels an Alexander or Napoleon to sacrifice millions of lives in his attempt to unite the world under his sceptre"... "and we are so accustomed to the phenomena that most of us fail to realise how abjectly stupid and undesirable the historical mass behaviour of humanity actually is."

The genius of Darwin has its moments of conceit. "The world," he writes, "... appears as if it had long been preparing for the advent of man: and this in one sense is strictly true for he owes his birth to a long line of progenitors." But many animals have just as long a line of progenitors as humans, most even longer.

Scientists today explain that mankind evolved from a killer-ape who broke apart from his non-aggressive cousins and advanced his status through leading a predatory life. In order to hunt he developed the ability to walk upright and by using his hands he developed his brain.

There are many people who still believe that humanity was created by God. According to Archbishop Usher's *Kallander*, God created mankind on March 23, 4004 B.C. When proposing this date the Archbishop ignored the fact that the Egyptians had already devised a real calendar in 4241 B.C.

The following statement by Sherwood L. Washbourne reflects the absurd reality of this and many other attempts to formulate a definitive chronology of man's development: "The study of human evolution is a game rather than a science in the usual sense."

I will start with the presupposition that one can see in the final product both its origin and its stages of evolution.

Mankind is a singular species in nature with a number of unique peculiarities. Logically, therefore, the history of humanity should be a history of these peculiarities. It is also logical that a unique end-product must have had a unique beginning.

From the very beginning our ancestors must have been more prone to development than other animal species. What kind of animal can be prone to development? Surely only an under-developed animal. The human species must have started from a state of inferiority, a unique inferiority.

Darwin proclaims that *Natura non fecit saltum* (Nature does not jump). This is true, but nature can create abnormalities or novelties not by jumping forward but by slipping behind, by halting at an earlier stage of development.

The central distinguishing peculiarity of humans is that they are ready to perform sexual intercourse at any time, while other animals are only sexually aroused during the so-called mating season, scientifically described as the time of oestrus. What is more, humans are the only species obsessed with sexual pleasure.

In order to understand what really happened at the beginning we must address another peculiarity of mankind. Human beings are the only species lacking the instinct to reproduce. Humans reproduce by accident as a result of sexual pleasure.

The amount of contraceptives used and the number of abortions performed underlines this. Mankind, however conceited or proud it is, must concede that behind the conception of nearly every man on the planet there lay no noble instinct of reproduction, no generous feeling toward the species at large; instead there was merely sensual pleasure, often drink or drug induced, often obtained by money, deceit, or rape.

It is impossible to enumerate the people walking the streets of this world whose conception was an unwanted accident. Why were their foetal lives not terminated?

The answer lies in a consideration of some religious, legal, or moral prejudice, or simply on account of financial straits. Most accidentally pregnant women would laugh at the poetic conceit of Novalis, who saw in every child an *amour devenu visible*.

The absence of the instinct for reproduction engendered an absence of the instinct for the preservation of the species. An absence of the instinct for the preservation of the species explains the ease with which humans destroy each other, their indifference to the future of mankind.

Before looking at the origin of these human peculiarities, it is necessary understand what happened before mankind came into existence.

Scientists estimate the creation of the earth at about 4,500 million years ago. Approximately 4,000 million years ago the sea was formed; 500 million years later the first life forms appeared in the water in the form of single-celled algae and bacteria.

Life must have been brought about on our planet by a particularly temperate solar energy in conjunction with mild climactic changes. These conditions must have introduced certain molecules of inorganic matter into an unstable or precarious state of existence. The ultimate aim of instability and precariousness is to reach a reduced instability, a reduced precariousness. Life, in essence, rests in the agitation caused by instability and precariousness in search of a reduced instability, a reduced precariousness. The agitation of the molecules in inorganic matter gave rise to self-replication, started producing growth, initiated a transition from inorganic to organic matter. Irritation is the main source of growth in the living world. Growth inevitably means over-growth and over-growth means a decline which brings organic matter to its previous state of existence as inorganic matter. That life consists of instability in search of a reduced instability and reduced discomfort can be deduced from the fact that the essential pillars of life, inherent in every living organism, are the organism's repair and immune systems. Their aim is to stem an organism's biological discomfort and to prevent its injurious acceleration. This aim must have been inspired and initiated by the basic tendency of living matter to reduce its instability to the minimal level.

We can calculate the appearance of the first oxygen-breathing animals to approximately 900 million years ago, and the development of various species of fish at 400-600 million years ago. 430 million years ago land plants began to grow. Between 300-400 million years ago the amphibians, reptiles and insects started to evolve. Approximately 230 million years ago the dinosaurs appeared, followed by mammals and, some 30 million years later, birds.

Around 70-75 million years ago, the ancestors of our ancestors, small rat-like insectivores, left their perilous life on ground level for a safer one in the trees. In these small, tree-shrew type of mammals who chose an arboreal environment for reasons of safety, is the origin of the primate. The first page of the history of mankind opens in the woodlands of East Africa.

Most scientists calculate the appearance of monkeys and apes to be about 40 million years ago. They explain that the oldest manlike primate, *Ramapithecus*, lived in Africa about 10 million years ago. *Australopithecus* lived in Africa approximately 1½ million years ago and is considered to be our direct ancestor. Approximately one million years ago we meet the so-called *homo habilis*, the tool user or tool-maker, and at the same time humans started walking upright. About 130,000 years ago Neanderthal man was to be found in North Africa and Europe, but 20,000 years ago they dwindled with the appearance of *homo sapiens*.

This is a general picture, as presented by most scientists today, of the evolution of mankind until the arrival of *homo sapiens*.

The popular writer Desmond Morris describes this important era of so many million years ago in his book *The Naked Ape* in just two sentences: "The ancestors of the only other surviving ape—the naked ape—struck out, left the forest, and threw themselves into competition with the already efficiently adapted ground dwellers. It was a risky business but in terms of evolutionary success it paid dividends."

Carlton S. Coon, in his book *The History of Man*, compresses the most vital events in mankind's past into the following three sentences: "From some kind of a Miocene ape probably living in Africa, both living apes and man are descended. The apes' ancestors, after a trial period on the ground, swung back into the trees. Ours stayed below, rose onto their hind-legs, made tools, walked, talked and became hunters."

In other words, humans and apes decided one day to come down from the trees, to leave an environment ideal from the point of view of food and security. Then the apes, the less advanced animals, returned to the trees and lived happily ever after, while the more advanced humans walked out of their natural paradise into the hell of the savannah.

So our human ancestors found themselves in the savannah with no natural specialisation and a brain not more than a third of the size of a gorilla's, i.e. one seventh the capacity of that of modern man: small and fragile of stature (about 2½ feet tall) they were to live and compete with highly specialised and dangerous predators. If we view their situation in terms of natural logic, we conclude that our human ancestors could only have made this step if they were urged on by a strange desire to commit collective suicide. We must remember that originally the ancestors of primates fled from the dangerous ground into the safety of the trees, a wise step which must have left a trace on their brain. No animal will ever leave a safe environment for a dangerous one, especially when it has an in-built and atavistic fear of danger. The atavistic fear of falling is occasionally recurrent in our nightmares.

Scientists do not explain why our human ancestors separated from the apes' ancestors and initiated their own evolution. Nor have they ever explained what the anomaly was which affected our ancestors and launched the human species, an anomaly which must have been present from the start, and one on which depends our separation from the other apes, and our uniqueness.

What follows is my explanation of the beginning of the human species.

In the Eocene epoch (36-58 million years ago) there was already a distinction between the anthropoids and the prosimians. The former were human-like primates, the latter included the ancestors of lemurs, tarsiers, and tree shrews.

Our ancestors lived in company with the ancestors of apes. The law of natural and sexual selection was in operation. From their former existence as lower mammals, the primates inherited a hierarchical system of organisation. In this hierarchical society the female assumes the rank of the male she is copulating with, her aim being to copulate with the highest rank possible. Primates only copulate when the female is in heat. During this period the female develops distinct signs which arouse the instinct of reproduction in the males. This was the system which human ancestors and ape ancestors respected.

When reaching sexual puberty, still dominated by an infantile mentality, some of the male apes must have become neotenous, neoteny meaning the retention of infantile or juvenile characteristics in adulthood. These apes must have retained their infantile state of sexual puberty for the rest of their lives. Man became an infantile adult. Even today, we often carry infantile traits into old age. Even today, an adult human looks more like an infant ape than an adult ape.

This sexual play, coupled with another infantile inclination, the imitation of the mature members of the community, became an abnormal novelty: our distant male ancestors started sexually assaulting females when they were not in a state of oestrus.

Amongst mature apes copulation took place mainly when the female was in heat, when her ovulation, proclaiming her readiness to conceive, produced visual or odoriferous signs which stimulated the male's sexual arousal.

Behaving in adulthood like infants, our ancestors must have irritated the other apes of the community. Soon our distant father's ancestors must have been chased out of the community by the mature apes.

Chased out of the community, our forebears formed gangs which followed the main community at a respectful distance.

At a certain stage, in the main community of apes, our mothers' ancestors must have started appearing. These females reached adulthood without ever developing a biological mechanism concerned with oestrus, the important phase of reproductive readiness.

In their immature state of sexual development, they also started imitating the mature female apes, trying to copulate with males without being able to stimulate them sexually: in the absence of the outward signals of oestrus, the males were not sexually aroused. This persistent imitation of the mature female apes in heat began to irritate the mature apes and resulted in the immature females being chased out of the main community. Their only option was to join the gang comprised of our fathers' ancestors, who were preoccupied with sex, and form a new community.

In these new communities love-making became the main play-activity.

This activity had no natural rules or patterns; it was performed day and night, in all circumstances and in all climatic conditions.

Our ancestors' frustrated and fear-induced exclusion contributed a great deal to their desire to make love. As their feelings of anxiety and fear increased along with their sense of exclusion, they sought refuge in calming proximity, togetherness, closer bodily contact, a reassuring clinging to each other. This encouraged cuddling, hugging, caressing and kissing and greater intimacy in general.

Cuddling, hugging, caressing and kissing also provided a protective sense of belonging which in turn inspired playfulness and sex-inducing sexual foreplay. Even today, the protective instinct inspires physical attraction.

So, we can still see people in closer togetherness and greater intimacy when they have been displaced, when they are chased or persecuted, when they have been excluded, when they are a minority in a hostile community or when they are ghettoised by economic misery.

It was for two different reasons that our male and female ancestors discovered pleasure in sexual intercourse. Success in performing the sexual act was an achievement, a fulfilment for an infantile male adult: it made him grown-up, a mature being. This feeling must have prompted the release in our ancestor's brain of pleasure-causing neurotransmitters such as oxytocins and endorphines which provoked an ecstatic or an orgasmic pleasure. This must have been so, for the simple reason that it is valid even today with modern man. That man's supreme pleasure consists even today in his achievements can be seen not only in his love-making but in other fields and even in games: when a man scores a goal in an important football match or when he wins a gold medal in the Olympic games, he enters into an ecstatic or orgasmic state.

Touching, hugging, cuddling or caressing their partners, our maternal ancestors must have felt satisfaction from the release in their brains of the same opiate-like pleasure-producing neuro-transmitters. Modern women seem to be no different in this respect to their maternal ancestors.

Being more developed and more mature than a man, a woman is less interested and less excited by achievement or by success than a man is. That is, perhaps, why she experiences less ecstasy and less orgasmic pleasure than men do.

A man's orgasm can leave him empty, bored, lonely, exhausted or lost. A football team scoring a goal can feel vulnerable to being scored against soon after they themselves have scored.

A woman who considers sexual intercourse as an achievement, a woman who is attracted by success or conquest, in other words a woman who *imitates* a man, can reach the same orgasmic pleasure and the same post-orgasmic emptiness and loneliness as her successful male counterpart.

These promiscuous sexual relationships enjoyed by our ancestors resulted in the accidental production of offspring. Even today, after millions of years of evolution, our reproduction is more of a miracle than a rule or a regular natural process. Even when our females' eggs are fertilised, approximately 70% of them lead to a miscarriage. Human child-birth is more difficult and more dangerous, for both the mother and the child, than that of any other species.

Children of undeveloped parents can only be undeveloped children. They, in fact, were the most helpless offspring of any species. They needed care, nursing, feeding, teaching and protection far longer than the young of any other species. The life-span of our ancestors was around thirty years, nearly half of which was spent in helpless infancy.

Our ancestors' pursuit of sexual pleasure must have inspired our species' main characteristic: the pursuit of pleasure. Our ancestors soon became obsessed with pleasure. In fact, we continue to behave today in the same way in which they must have behaved. This should not come as surprise, since the pursuit of pleasure leads to passion, and passion tends to perpetuate the infantile mentality. The passion for pleasure prevents our brain evolving toward maturity. In fact, an infantile mentality does not seek a state of maturity. Instead it seeks to avoid it, since maturity implies reason and rationality and these are not the stuff of fun and excitement.

Fortunately for our species our maternal ancestors, through becoming pregnant, acquired a maternal nature and with it a sense of responsibility, an instinct for caring, nursing, home-making and protection of the community. This is scientifically manifested in the important hormonal and physiological changes that take place in the bodies and brains of a pregnant women to this day.

Yet notwithstanding the primeval protective instinct, human reproduction continued not as a result of a natural drive but as an accidental side-effect of the pursuit of sexual pleasure. Later the tendency towards pleasure for its own sake led to the use of contraceptives as a means to achieve pleasure without the attendant responsibility of procreation.

It was in the birth of the offspring of two original sexually-undeveloped apes that the future of mankind was mapped out. The female soon realised that the best way to keep a male with her and her offspring was to tolerate his pursuit of pleasure and to provide him with what was, in effect, 'motherly' attention and care. Even today man feels attracted to a woman who cares for him and who loves him and is more likely to remain attached to such a woman. In his infancy, or infantile mentality, a man loves himself primarily, and those who love themselves in turn crave to be loved.

As man was in his infancy, it seemed logical and natural that the first early human communities should be dominated by mothers: they were the custodians of the essential survival-values of our social species, instincts like loving, caring, nursing, nurturing, sharing, co-operation, health care and the tending of the sick and wounded. Women who attained old age must have enjoyed particular respect and authority since their knowledge, experience and memory were invaluable to the community.

When analysing our present sexual behaviour, we might agree that my theory concerning our ancestors' sexual behaviour is plausible. Our present obsession with sex is evidence that we are, particularly, we men, still in neoteny. Our immature ancestors developed their sexual habits through the imitation of other, more mature, apes. In fact, we have never developed a definitive innate technique for love-making. We have to learn it, and we have developed a variety of techniques; some of them are even bizarre and perverse. Manuals setting out to teach us various modes of love-making and the best ways of achieving orgasm are very popular.

The human being is still the only animal who observes no natural rules as far as love-making is concerned. We have sex day and night, in all seasons, in all climates and in all environmental conditions. Most other animals have specific periods in which they copulate. This period is determined by the females of the species, by their readiness to conceive. In nature, in fact, the drive towards copulation is instigated primarily by females. Males are sexually stimulated by audible, visual or olfactory signals given off by the aroused females. The male, who as we have seen is more obsessed by sex, seizes the initiative. What in primeval times were natural signals have now given way to artificial ones:

make-up, scent and pretty clothes bear no relation to oestrus and are largely responsible for our round-the-clock readiness to engage in the sexual act.

So it is that our females either fail to present signs of their readiness to become pregnant or, in his obsession for sex, these signs are imperceptible to men. What is unique to man as an animal is that he will even ignore a woman's most evident repulsion of his advances, even to the extent of rape, such is his overwhelming obsession with sex.

And what is even more curious is the fact that our females do not instinctively realise when they are receptive to the fertilisation of the egg.

Our sexual relationships seem not to have changed for millions of years: they are still mainly a question of playfulness, excitement and fun. In fact, the more jovial and playful a person is, the better his or her sense of humour is: he or she is more attractive and possesses greater sex appeal. Even some outright defects like lisping, for example, which gives an impression of playfulness, can be seductive. Often, during love-making, we tend to use endearing and playful baby-talk. We tend to be attracted by baby faces.

In other species a mature male would seldom try to copulate with a female if she was not on heat. This is a crucial sign of maturity, even though it has to be said that in other species the infants and adolescents may try, in the spirit of imitative playfulness, to copulate with females who are not sexually receptive.

Men have sex with women even when they are already pregnant, after their menopause, when they have had hysterectomy, and even, in the case of necrophiliacs, when they are dead. Man is the only animal known to have sex with members of *other* species. We are also the only species to have sex with children.

We are proud and consider ourselves superior to other species, proud to be able to spend so much of our time and energy in the games of pleasure, excitement and fun which both precede and sustain our love-making. If we had spent the time and energy that we have spent on love-making in trying to improve our standard of living, we might have been able to create a paradise on our planet.

We are the only species obsessed by aphrodisiacs, supposedly designed to increase our sexual prowess and potency. The success of the male potency drug Viagra is new evidence of man's obsession with sex.

That sexual intercourse is more a playful performance than an innate drive aiming at procreation is evident in the language we use to describe the act. In English we call it love-making, in French *faire l'amour*, in German *Liebe machen* and in Italian *fare a l'amore*.

In fact, the Catholic Church would be right to consider abortion as a sin if our copulation was dictated by a natural drive and not by the pursuit of pleasure, excitement or fun.

We often behave like spoilt children: we will pray for sex, beg for it, to risk prison for it, pay for it or even kill for it.

With other animals sexual intercourse is often a matter-of-fact activity. After sexual intercourse most

17

other animals simply continue their day-to-day life, hunting, sleeping, grooming, etc. Since our love-making is an exciting game it reaches climactic orgasm which in turn depletes our physical and mental resources. In this sense it resembles all other games: tremendous exhilaration followed by great exhaustion. It can leave us tired, lonely, sad, empty or depressed. It has evolved into an gruelling trial rather than a natural performance, to the extent that some people suffer heart failure, succumb to strokes or even die during sexual intercourse.

What is more, during sexual activity, the efficiency of our senses, of our perceptions and of our very reasoning can be reduced to a dangerously low level.

That men consider sex as play, excitement or fun can best be seen in the pantheon of deviation, perversion, transvestism, sodomy, sexual wickedness and voyeurism. Basic infantile nastiness can transform sex into sadistic and masochistic pleasure, excitement and fun.

Man's obsession with sex can also be explained by another consideration. In his neoteny, his prolonged retention of the immature characteristics in adulthood, man sees in woman a maternal protective force: he feels safe in the embrace of her intimacy and this in turn perpetuates his infancy.

We practice kissing more than any other species because it reminds us of suckling at the breast. In fact, it must have been suckling, the most pleasurable activity of early infancy, which originally inspired the act of kissing. When we kiss, we tend to close our eyes like suckling babies. The more fragile and the more infantile a man is, the more he enjoys kissing.

Could it not be, after all, that many other new species have evolved from less mature by-products of their own species? Less developed individuals are more flexible and more open to change. The fit individuals of a species tend to perpetuate themselves.

That man is an undeveloped animal can also be deduced from the fact that when he discovered his mind and its capacity to create wishful fantasies, his mind in turn created a wishful self, a pretentious ego.

Darwin realised that humans evolved from inferior apes, but he did not dare to emphasise the fact. It was already considered deeply offensive in his time to suggest that man and ape had common ancestors. "Hence it might have been an immense advantage to man to have sprung from comparatively weak creatures," he wrote in *The Descent of Man*. "We have seen" he wrote in the same book, "…that man bears in his bodily structure clear traces of his descent from some lower form."

That our ancestors must have regressed to neoteny can be deduced by the fact that we are still in the same phase of development.

The immaturity of our brain can be easily noticed in our distinct lack of strong and uniform patterns of behaviour. We are also slower and more unpredictable than other adult apes when confronted by sudden threat or danger. In contrast to other higher primates, we are capable of entering into a stupor or fainting in the face of unexpected peril.

Being in an infantile phase of the development, our brain is more plastic, less well organised than that of the great apes. We are less reliable, less predictable, less rational and less responsible in our attitudes and behaviour than chimpanzees or gorillas.

Our neoteny can be confirmed mainly, however, by the evidence that our canine teeth, our body hair, our shoulder muscles and our jaws had remained undeveloped in comparison to our ape cousins. The bones of our children are more cartilaginous than those of the young apes. While we are mainly right or left-handed, great apes are mostly ambidextrous.

We have also less developed and less efficient natural defences than the other higher primates. In fact, before we discovered hygiene, healthier nutrition, immunisation or prevention of diseases, our average life-span was lower than that of the great apes. Many viruses which are harmful or deadly to us are kept under control by the more efficient immune systems of the higher primates.

Further evidence of our poor development can be found on examination of the human childbirth cycle. Despite a much longer gestation period than that of other primates we produce far frailer, less self-sufficient and well-developed infants, and our cousin apes reach adulthood far more quickly than we do.

Could it be that because of our underdevelopment we are the least graceful, least elegant, most awkward species in nature?

It is interesting to notice also that chimpanzees and gorillas have more chromosomes, the carriers of genetic material, than our species Scientists also inform us that there is only two percent in the genetic difference between our ape cousins and us. They do not explain, however, if this difference is to our or to the apes' advantage.

Our neoteny could, perhaps, explain why we are more open than other primates to the harmful mutations in our genes? Could this explain our typical human ailments? Could it be that our strong predisposition to genetic mutations is the major cause of that noticeable individual differences in our species? In fact, we are far more different from each other than the individuals of any other species.

Given our species' strong predisposition for genetic mutation one can ask if it is wise to allow genetic modifications of our food?

Our species' neoteny might have helped us to discover speech and to enrich our verbal and sign languages.

Our confused patterns of behaviour, our slow responses to external stimuli, our high impressionability, due to our underdevelopment, must have made our ancestors vulnerable to a vast range of enduring and ineradicable emotions. Expressions of these emotions soon became communications. In fact, it is these emotions which provide the energy needed to manifest speech and body or sign languages.

21

As most emotions in our world and in the animal world seem to be caused by biological irritation, discomfort or suffering, most the languages in nature seem to consist of laments or complaints. Contentment tends to be more tranquil and, significantly, silent phenomenon.

That our language is expressing mainly our discontent can be seen also in the fact that a great deal of it, throughout space and time, consists of swearing and blaspheming, offensive oaths or curses.

With a vast range of communications, our ancestors discovered a deeper togetherness and a closer sense of bonding, so needed by our undeveloped species. In fact, even swearing often tends to create a lighter atmosphere, a more agreeable intimacy, a common ground.

Life in the Savannah

In the Miocene epoch, between 13 and 26 million years ago, a major event occurred in the evolution of mankind: a change of climate took place. A deterioration of the climate transformed a great part of the woodlands into the savannah. This new environment caused major changes in the life of primates. Previously the vastness of the woodlands and the abundance of food had enabled the primates to tolerate one another. There was no territorial enmity. When available space diminished, non-toleration of humans by apes set in. Reduction in food and space created an awareness of territory, and this awareness bred intolerance.

Here I stress that the instinct for reproduction in animals has its own self-checking brake mechanism which operates whenever the environment cannot support an increase in population. In the humans of that far-off era reproduction was not controlled by instinct but by a desire for pleasure, and there was no brake mechanism: the same is true today. Instead, for some abnormal or typically human reason, humans tend to copulate more avidly in times of crisis and thereby increase their population.

Soon the conflict between apes and humans became a fully strategised war to the death. Humans, in their ever increasing numbers, were creating a serious threat to the survival of the apes. The apes, however, were better equipped for battle. They had abided by the laws of natural selection, which decree that the fittest will survive for reproduction. And they had developed their most ferocious weapons, canine teeth. Our human ancestors no longer possessed these aggressive tools, and the apes started to chase them out to the open savannah.

So it was that about 16 million years ago our ancestors were evicted from their woodland paradise into the hell of the African savannah. This traumatic experience, this ejection from paradise, has remained a scar in mankind's brain for eternity.

Most anthropologists assert that human beings, the brightest of existing animals, set forth one fine day out of a natural, ideal environment, where there was food in abundance and little danger, and chose to start a new life in the hell of the savannah, with limited food, no safe shelter, and an environment filled with dangerous predators, particularly deadly serpents, which humans and apes feared and still fear hysterically. This theory does not bear logical scrutiny, since no animal will voluntarily leave a good environment for a bad one. It cannot be explained by the reasoning of our ancestors — a reasoning, if it can be called such, dictated by sexual pleasure, a pleasure more safely performed in the woodlands than in the perilous savannah. No, the theory is buoyed up by modern human logic, which is based on conceit. Humans, considering themselves to be the most advanced species in nature, must proclaim that the departure from paradise was a progressive step made by their ancestors, determined by free will, another of our illusions.

As we shall see, the human brain only began to expand quickly about a million and a half years ago, and only became capable of abstract thought and conceit about 30,000 years ago. Our ancestors, therefore, could not have been so conceited as to be so irrational.

I think that the most compelling evidence that humans were evicted from the woodlands into the savannah is the fact that the human mind, when it started to speculate, created the idea of a lost paradise. The mind's creative ability is embedded in the residual scars of past experiences, scored deep into the brain of the species.

Mankind is the product neither of fallen angels nor of elevated apes. Mankind consists of fallen apes. The Bible is more accurate than science in its explanation of the origin of human life. Our human ancestors, our Adams and Eves, *were* evicted from Paradise. The only difference is that historically they were not evicted by Almighty God, but by fitter apes.

We know that one ancestor, *Australopithecus*, who lived approximately a million and a half years ago, was about four feet tall with a brain capacity of between 435-600 cubic centimetres. *Australopithecus'* forebears, when they first came to the savannah about 16 million years ago, could not have been more than two feet six inches tall, with a brain a third of the size of gorilla's today. This frightened, feeble and underdeveloped creature, by nature defenceless, and with no special talent or aptitude for life in his new environment, makes a laughingstock of the theories put forward by the anthropologists who aim to glorify our ancestors.

Human frailty and lack of specialisation became, in fact, the main props for survival in the savannah. If humans had been stronger and more specialised, like any other specialised animal in a new environment, they would have exhausted the potential of their specialisation to an extreme degree, and become extinct as a species. Specialisation is distinguished by perseverance, which in an unnatural environment can be fatal. An underdeveloped animal is open to changes, to further development.

By the close of the Miocene era, the basic human stock was living in the savannah, about to confront the Pliocene epoch, which lasted from 13 million years ago until 1 million years ago. This was the most testing time in the history of mankind—an era of climactic deterioration and droughts which transformed Africa into a graveyard for many species. This climatically aggressive situation left a deep scar on the old human brain, a scar which influenced the mind in its creation of the first idea of Hell. It says, in the Sumerian epic *Inanna's Journey to Hell*, that "Here is no water but only rock, rock and no water and the sandy road..." In *The Epic of Gilgamesh* the writer states that "Hell is frightening because of its sandy dust" and because it is "the dead land..."; it is "the river which has no water." In *Inanna's Journey to Hell*, when the goddess Inanna is revived from the dead and is ready to return to the land of the living, one of the judges of the nether world says: "Who has ever returned out of Hell unharmed?"

Our male ancestors started life in the savannah by regressing even more to neoteny, paedomorphosis, or to the Peter Pan evolutionary phase. Even the modern male, with all his intricately-developed mental capacities, tends, after failure, to revert to infancy. It is as if he wants to become a child again in order to grow up and prepare himself all the better for a new life.

Many scientists agree that in the early stages there was an infantile phase in the savannah. Their belief is that it was mankind, both men and women, who went through this phase. It is my contention that if this were the case, we would not exist as a species today. It would have contradicted the elementary laws of nature if a mother had regressed to infancy. A woman matures as a result of pregnancy and acquires a feeling of responsibility toward the species.

Many scientists simply devote a mere paragraph to the notion that mankind faced a phase of infancy, often dedicating the next paragraph to the domineering and glorious male hunter, the generous and brave provider of food for his family. Infancy means dependence; the state of infancy requires a mother, a mother's guidance. The instinct system in infancy is confused and unreliable, even in animals that possess a strong system of built-in reactions. The infant has to learn; the human infant even more so owing to his lack of specialisation. The mother teaches him. Humans achieved a cultural instead of a genetic transmission of behaviour patterns.

It is evident, therefore, that mankind in the savannah must have continued to be dominated by women.

What does reverting to infancy mean in practical terms? Infancy is about play. The human species owes its survival to play. Play explores the environment and is guided by curiosity. It brought to mankind its most distinguishing characteristic, opportunism.

The permanent desire for new experiences, paramount in infancy, became through a greater sense of enquiry part of the nature of human males. It will always flourish — for men feel that a new environment is not their natural one, but merely temporary. They feel that through exploration they will sooner or later find their lost paradise.

When he discovered the creativity of his mind, man invented Paradises, Kingdoms of Heaven, Kingdoms of God and Utopias.

Exploratory play — an elastic, incomplete, and experimental activity — is what men lean on to discover the best way of adapting to their new environment. Play is after all the only activity which suits the nature of an incomplete animal such as man — an animal suffering from a sense of inadequacy. Play, which includes imitation, became man's only specialisation, because it was the easiest way of adjusting to the variety of life in the new environment.

The life of opportunism, discovered through play, had the important effect of perpetuating man's infancy for several million years. A life of exploratory play is filled with danger and accidents. Nature, though, is not playful. Whenever man in the course of play encountered the sense of determination which characterises a mature situation, or was challenged by the result of a specialisation, he would flee and run to his mother for protection, for encouragement: in simple terms he would scamper back for a new dose of infancy.

So it was that the human female found herself in the savannah with her maternal care instinct greatly increased as a result of her male's increasingly single-minded regression to infancy.

In the savannah our mother's ancestors soon developed, probably through their maternal instinct, a gift which has always remained with them: their easy adaptability to new circumstances. Woman was assisted in her easy adaptability by her enhanced instinct for imitation, her respect of nature and her intuition.

Man, meanwhile, remained in his childhood phase throughout the Pliocene epoch, which lasted for nearly 12 million years.

Life in the savannah meant organisation and a division of labour. The female-dominated human group would find a protected spot in a cave where a home would be formed and where life would be organised. Mothers with offspring would stay in caves while the males went off in search of food. Man started life in the savannah by food gathering, bringing it back to his group in exchange for protection and sexual pleasure. Over the past millions of years, man has not changed. If man had not been in an infantile state, and therefore dependent on women, and if his instinct for survival had been stronger than his need for biological comfort and pleasure, he would never have returned with the food. In the case of other primates, the group follows the males.

Even today, most men only share what is theirs in return for comfort and security. From the earliest legal codes to the present there has been a steady development of judicial structures which impose duties on a man as far as his family's maintenance is concerned.

Humans became omnivores, the males through exploration, the females through adaptation. Man, behaving just as baby does, put everything that looked new and interesting straight into his mouth — eggs, small birds and young animals, insects, roots, plants, berries, meat from carcasses left by other animals. By this method of trial and error he evolved the peculiarly eclectic human approach to diet that is still so much in evidence today. Judging by the quantity of animal bones and skulls found alongside early human remains, marrow and brains were clearly part of early man's habitual fare.

Most scientists claim that men, as soon as they came to the savannah, started hunting and providing food for their families. Our overwhelming conceit has produced this romantic fallacy. Later I shall show how man is the only animal capable of self-deception.

One can make nonsense of this question of hunting by asking a simple question. How and with what weapon could man have hunted? As the spear was not invented until 30,000 years ago and the bow and arrow not until 12 to 15,000 years ago, I fail to see how this was possible. Man started his life in the savannah with neither natural nor artificial weapons. Hunting is an activity either dictated by instinct or by abstract thought, and man, prior to *homo sapiens*, had neither. In the course of his sheltered woodland life he had lost that keen sense of smell inherited from the lower mammals. Though in his arboreal existence his eyesight had become strongly developed, the conditions in the savannah dictated that he was not so much concerned with seeing than with the important business of *not being seen*.

The human males formed packs in order to gather food. They might have started this in imitation of other animals, particularly hyenas. They searched for carcasses killed by some lone predator, whom they would chase from his feast by dint of sheer numbers. If this ploy failed they would seek out a wounded animal.

The confidence implicit in safety in numbers is yet another scar on the old brain that has fed and programmed the new one, influencing man's creative thoughts. It is reasonable to suppose that overpopulation can be ascribed, at least partly, to this ingrained belief.

The Human Brain

Towards the beginning of the Pleistocene age, humans were about four feet tall. The capacity of their brain, however, was still only about 600 cubic centimetres, two and a half times smaller than the brain of *homo sapiens*.

The Pleistocene age saw two new phenomena. The first was an extraordinary increase in the size of the human brain and the second was that humans began to walk upright.

In the period from approximately one and a half million years ago to about 200,000 years ago, the human brain increased two and a half times in volume.

Most scientists claim that the human brain grew as man used his hands more effectively as a result of bipedal locomotion. This theory has it that humans, in order to develop their brains, needed to free their hands. To free their hands they forced themselves to stand upright, to walk on two feet, and to place themselves in a thoroughly unstable and unnatural position. This theory does not explain why after so many millions of years mankind suddenly discovered manual dexterity in a flash. If bipedalism and manual ability had helped humans to increase their brain efficiency, then the dinosaur would still be with us and the kangaroo would have a much bigger brain. Extending this theory, our cousins the apes should have bigger brains if their manual dexterity is anything to go by.

What, then, could have been the reason for the increase in size of the human brain?

Irritations and frustrations seem to be the main causes of growth in the living world. This is in tune with the supreme law of life in which irritation, instability or discomfort in cells, organs or organisms is remedied or diminished through growth. The following exposition endeavours to show the relationship between the brain and its fears, irritations and frustrations.

In reality we have three brains which are interconnected. They can also behave independently sometimes. We have the old brain representing our reptilian past, we have the limbic brain representing our mammalian legacy and we have the new brain, which is particularly developed in our species and which we inherit from our primate ancestors.

Each of these brains has its own anatomy, its own chemistry, its own rationality and reasoning, its own values and its own attitudes and behaviour.

Frustrations, irritations and fears played the leading roles in the expansion of our brains.

The purpose of the development of any new brain was to diminish the frustration, irritation or fear to which the previous one had fallen victim.

By developing learning and memory the limbic brain helps to placate or to reduce the frustrations, irritations or fears of the reptilian brain, which, since it operates on the here-and-now, short-term basis, has difficulty in grasping that many things which at first sight are frightening are not dangerous in reality.

By developing sociability, gregariousness and the group life, the limbic brain counteracts the reptilian brain's fears of loneliness.

Vocalisation and hearing, important faculties of the limbic brain, assist in communication: through them the group can be mobilised against danger of any sort.

On the superior level of the limbic brain, which is particularly well developed in our species, we developed typically humane characteristics such as sympathy, empathy, friendship, sharing, and an instinct impelling us to take care of the old. All of these eliminate or at least reduce a great deal of our frustrations, irritations or fears.

Along with our big new brain we acquired the ability to *reason* ourselves out of frustrations, irritations or fears and to de-dramatise threatening events and thus lessen the sense of panic that increasingly beset us in our new environment.

The relationship between fear and our three brains is evident whenever our new brain is unable to reason us out of a fear. Whenever our limbic brain is unable to cope with fear, we return to the reasoning and values of our reptilian brain. The stronger the fear or irritation, the more prominent the reptilian legacy.

Whenever our limbic brain and our new brain are reduced in efficiency by alcohol, drugs, stress, disease or old age, our reasoning, attitudes and behaviour become influenced by our reptilian inheritance. Infants who have not developed or who never will develop a limbic brain in tandem with the new brain will inevitably fall back on reptilian or autistic modes of behaviour.

The main behavioural modes of the reptilian brain are: selfishness, self-centredness, ruthlessness, treachery, viciousness, suspicion, callousness, cruelty, cold-bloodedness, rigidity, stubbornness, perseverance, individual rivalry and competition, individual assertiveness, territoriality, reserve, impatience, intolerance, aggression and violence, here-and-now gratification.

The mammalian or the limbic brain evolved around the reptilian brain under the pressure of the reptilian brain's frustrations, irritations and fears.

The limbic brain's main abilities are: sociability, gregariousness, family and group life, curiosity and exploration, learning and memory, vocalisation and hearing, communication and play, touch and intimacy. All of these reduce frustration, irritation and fear.

With its centres for communication and language, the limbic brain helps to sustain a calming intimacy and togetherness among the members of a community. When we are mortally afraid or on drugs we lose the efficiency of our limbic system and therefore return to reptilian body language.

The limbic brain's sociability atones for reptilian violence, and ruthless individual competitiveness or rivalry are replaced by individual co-operation, by a community life. Intraspecific fights amongst mammalians are far less violent than reptilian ones. The mammalian territoriality is a reptilian legacy which comes into prominence when we are threatened or frightened.

Every experience is accompanied by an emotion. Each experience, in fact, is registered in our memory pool on its own specific frequency of emotional energy. For example, episodes that occurred in a state of drunkenness will be remembered more vividly on the same level of drunkenness.

When the limbic brain loses its efficiency owing to stress or fear this causes the reptilian brain to come to the forefront and we can at that point lose our memory.

Mammalians, primates and humans are born with the neuronal mass of the limbic brain, but in order to have an efficient limbic system, this mass of limbic cells has to be formed and structured. This is done mainly through experience and learning in early infancy. If the limbic brain has not developed its needs for sociability and togetherness, for communication and speech, for caring and mothering, in infancy, it will have difficulty in developing them later.

A child is born a selfish, self-centred, ruthless reptilian and he will remain dominated by the mechanism of the reptilian brain for the rest of his life if in his early infancy his mother and the community have not taught him the valuable potential of the limbic brain.

Because of the fragility, vulnerability and fear to which we are prey, we developed superior layers of limbic brain which can be called 'the humane brain', with its capacity and in-built desire for dealing with typically humane feelings and behaviour such as: friendship, affection, pity, sympathy, empathy, compassion, gentleness, love, sharing and caring for the sick and the old.

This theory can be proved by the fact that when our limbic brain is damaged or shaken in its efficiency by drugs or fear, our humaneness is reduced or even disappears.

When our limbic brain is in perfect harmony with the new brain we then develop what can be considered the supreme human values: a sense of gratitude and a sense of humour.

When we develop the humane brain's centre for loving, we develop a need for loving and contentment. Our loving instinct demands reciprocation, and so there is no love unaccompanied by a need for loving. In fulfilling this need we feel we have achieved a safer existence.

Stimulating one of the centres or needs of the humane brain triggers off the activities of other centres in the same brain. Loving people become more sympathetic and friendly and ready to help those in need. Triggering off the centre of the sense of humour, easily triggers off centres such as generosity. Caring for a pet, for example, makes us more humane in relation to humans and nature.

The main function of our neo-cortex, or new brain, was to help us reason ourselves out of fear and frustrations and, by analysing fearful events, render them less traumatic.

Many scientists agree that even the other primates use their new brains in more complex problem-solving and that the difference between them and us is only a matter of degree as far as mental activity is concerned.

We consider other primates inferior, however, as they do not have the capacity for abstract, speculative or wishful thought.

In my view, the most significant difference between other primates and us is that while other primates' new brains work in harmony and collaboration with their limbic brains, our new brain often operates *independently* from our limbic brain, sometimes even against it.

In its independent activity our new brain developed such phantasmagoria as abstraction, fantasies, illusions, day-dreams and wishful beliefs.

At birth our new brain is poorly harmonised with our limbic system. In fact, children's independent activity or the fantasising of the new brain is a well-known phenomenon.

Inspired by an innate tendency towards playfulness, the new brain starts playing games with itself, creating an unreal world of imagination. On growing up, which implies development of the organisation of our limbic brain, the brain reflecting reality, this independent activity of the new brain becomes limited.

What could have been the main cause of the increase in our male ancestors' fears, irritations or frustrations, which must have been the major instigation for the big increase in brain-size?

My answer to this question is our male ancestors' *awareness*: awareness of their shortcomings, of their inadequacies, of their insufficiencies. Only an undeveloped animal could have been able to become aware of itself and of its failings.

This awareness of inadequacies and insufficiencies, of shortcomings and discomforts, must have increased our paternal ancestors' fears, irritations and frustrations and therefore these exerted permanent pressure on their central nervous systems and were the major cause of the increase of brain-size.

The females, meanwhile had a better organised and more mature limbic brain. They took in their stride the growing fears and frustrations of their male partners and offspring and were better able to cope with day to day life. They were comparatively tranquil and therefore their brains did not increase so dramatically in volume as those of their menfolk, because the vital stimulus of irritation and fear did not figure so much in their development.

Homo Erectus

Another great mystery in the history of mankind is why and how humans became bipedal creatures. Some writers explain this change as a need to free the hands to use tools and weapons. But I believe that man did not begin to use weapons until long after he had been standing upright, and that for a long time in his new posture man continued to be only a food gatherer. Tools, such as they were at the beginning of man's bipedalism, were used with equal dexterity by apes, who did not walk upright except on occasions. Thus, handling tools and weapons was not enough of an incitement for man to continue walking on two feet and standing upright, tiring things in themselves, at a time when he had no special need to use his hands.

If one accepted this explanation of the reasons behind upright posture, the human female would still be on all fours.

Some writers explain that man became upright because from this new posture he could spot his prey more easily. My answer to this is that it is more in keeping with natural logic that a highly vulnerable creature like man, with no offensive or defensive weapons, and lacking the speed of other animals, would have been much more worried about being seen by his predators than spotting his prey.

Improvisation, which is in the nature of an opportunist, is more difficult in a precarious bipedal position than on all fours.

What happened then? To understand what I am proposing, one must take into consideration the fact that the upright posture coincided with the increase in the volume, and therefore the weight, of the developing human brain. The phase towards gradual bipedalism followed the stages of the gradual increase in the size of the brain. It would seem that erect posture was attributable to extra weight in the cranium. Erect posture was not a choice of man but was forced on him. It was forced on him by this extra weight—approximately 800 grams—of the new brain.

It may seem unbelievable that such a small increase in weight could have produced such dramatic consequences. Had the extra weight been borne by a creature living without stress in an arboreal environment on all fours, it might never have figured as a major component of our development. But life in the savannah was exhausting, uncertain and dangerous, and the small weight began to seem like a huge millstone.

Let us look at familiar example. A pregnant woman always stands with her back arched, and wears flat shoes in order to keep her centre of gravity in the right place.

Labouring under the pressure of the increased weight of their brains, humans were forced to stand upright, balancing the head on the spinal cord.

Darwin was very close to agreeing with this explanation of the erect posture. "The gradually increasing weight of the brain and skull in man must have influenced the development of the supporting spinal column, more especially whilst he was becoming erect," he wrote in *The Descent of Man*. "In young persons whose heads have become fixed, either sideways or backwards, owing to disease, one of the two eyes has changed its position and the shape of the skull has been altered apparently by the pressure of the brain, in a new direction."

Scientists say that when a baby stops crawling and begins trying to walk on two feet, he is urged on by an instinct for bipedalism. Instincts, like God and genetics, are used as easy solutions to cover up ignorance of the real causes.

If bipedalism had been an instinct or an innate posture, then the human infant would not take so long to walk on two legs as he does. In the animal world the infant assumes the natural posture of its species a few days after its birth and, in some cases, after only a few hours. If bipedalism were an innate position for man, then humans would not so easily get tired while standing. Many other animals relax, rest, and even sleep in their natural standing-up position. We can deduce that the upright posture was an imposition rather than a benign natural development, since it has brought with it extra fatigue, discomfort, not to mention new illnesses such as curvature of the spine, kidney trouble, back pains, and varicose veins.

The Appearance of the Mind

Our mind is unique in nature. Our mind emanates from within our brain and is fashioned by the brain's abstract creations such as: illusions, fantasies, imagination, fiction, day-dreams, beliefs, revelations, myths, fairy-tales, hallucinations, self-deceptions, delusions and other kinds of irrationalities. Because it is irrational and supernatural, the complex world of the mind seems to escape genetic determinism and the Darwinian laws of evolution.

We spend most of our lives in the mind's irrational and supernatural world. We are proud of it, failing to consider the irony that we label people who take the irrational/supernatural aspect of things too seriously as bring 'mentally abnormal'.

At the same time we consider ourselves superior to other animals because we are able to live in the absurd world of the mind's fantasies. We often spend our entire lives playing silly games in which the mind invents a wishful belief on which it leans, using its imagination to justify an irrational belief. This is best seen in strong religious or ideological believers.

How did we discover our mind and how did we become dominated by it?

At a certain stage of our evolution we became aware of our brain. This awareness came when we started noticing its inadequacy to cope with our needs, needs of a fragile being in a state of neoteny. We tend to notice an organ when it hurts or when it causes some discomfort.

In his awareness of his brain man might have been helped by his upright posture. Due to the force of gravity, the upright posture might have brought some changes in the blood affluence to the brain, the brain which was the very cause of bipedalism. The sensitivity of the human brain to blood supply can be deduced by the fact that even a slight diminution in this supply, caused by drugs, malnutrition, altitude or thin air, rhythm of breathing, can create an increase in the dissociation of the mind from reality, increasing fantasising and hallucinating.

His use of narcotics and alcohol might have contributed to man's development of his mind.

When man became aware of his brain, he approached it with a predictable exhibition of the main characteristics of his infantile mentality, his wishfulness and his playfulness. In fact, our mind is a product mainly of our *playing* with our brain. The best evidence that the mind is created by an infantile mentality is that it functions with an infantile logic and an infantile reasoning, that it can be silly or funny, that it can be capricious, nasty, aggressive, cruel, violent, destructive, unscrupulous, irresponsible and amoral. All of these characteristics belong mainly to infancy or to an infantile mentality. All of these characteristics are also characteristics of our reptilian brain which comes into prominence when we are open to fear. Placing us in an absurd surrealistic world, our mind increases our feeling of precariousness, thus increasing our fears. Some of these characteristics, especially aggression, violence and cruelty, became permanent components of human behaviour and are the major driving forces in human history.

As the mind developed, man became an even more absurd animal. We are proud of our big brain, but just one stubborn abstract belief, one fixed idea, one absurd prejudice or one ridiculously self-flattering self-deception of our mind's imagination is enough to reduce or to eliminate the rational capacity of our brain; the bigger and stronger our mind, the less efficient is our brain's common-sense intelligence. As our mind-world is growing bigger and stronger, our future does not seem promising. By now, we are reaching a situation in which the strong-minded, those who are the most aggressive, are more successful and more appreciated than kind and intelligent people, a situation in which the mind has posited aggression as a virtue. This allows a mind-dominated culture to consider those who are guided by their reptilian brains to be measured as the fittest, and thus to dominate the world. What is even more ironic is the fact that unhealthy and fragile individuals tend to have stronger minds and to be more aggressive than healthy or wise people, dominated by their rational thus unaggressive brains.

Since they are more mature than men, women, and particularly mothers, are less attracted by the mind's surreal world.

As we have a great variety of shortcomings and frailties and an even greater variety of intensity of these shortcomings and frailties, we have a wide variety of human minds amongst us, a wide variety of beliefs, ideologies and prejudices.

Since the mind became dominant, our brain has become a tool of the mind that it created. It is in the nature of the creator to fall in love with its creation. The best example of the mind's domination of the brain can be found in the fact that the brain devotes all its potential to creating the most powerful and the most aggressive weapons to assist our mind's absurd religious or ideological beliefs in eliminating other people's absurd religious or ideological beliefs or to prevent their inception. Most of the cruellest wars, persecutions, exterminations and inquisitions were undertaken mainly in the name of religious or ideological beliefs.

With the appearance of the mind, man introduced another unique quality. Lacking innate patterns of behaviour, man started to replace them with the mind's created dogmas, rigid beliefs, stubborn prejudices or fanatical religious or ideological credos. This placed man among the most accident-prone of animals.

With the precarious world of the mind, our fears increased, and with the increase of these fears, our selfishness grew. In fact, the stronger a mind, the more selfish and self-centred it is.

The mind, at least in some cases, must have appeared around thirty thousand years ago. It is in this period that we see the first evidence of the arts, which are mainly the mind's creations.

It is curious to notice that the appearance of *homo sapiens* and his aggressive mind coincides with the disappearance of the pacific and unaggressive Neanderthal man.

What might have helped man in the development of the mind? The answer to this question might be: his dreams. Man's dreaming might have easily helped the discovery of his day-dreaming. In fact, dreaming is the earliest activity of our brain as it starts already when we are in the foetal stage. (Most probably it is the mother's emotions which are reflected in form of pictures in the foetal brain which form its dream sources.)

There is no big difference between dreams on the one hand and day-dreams, reveries, illusions, fiction, miracles, delusions, curious beliefs and fairy-tales, on the other.

As in dreams, the notion of time and space disappear in our day-dreams, replaced by the here-and-now.

Inspired as they are by wishfulness, our day-dreams are often a pleasurable activity, excitement and fun; and pleasure, excitement and fun are the main aspirations of the infantile mentality.

In fact, it is wishfulness, a phenomenon inherent in any incomplete or immature being, which is the major stimulus of day-dreaming. This wishful day-dreaming can best be seen in its creation of the mind's ego, this creation of the fragile infantile man who is unhappy with his real self. In fact, through our ego's pretensions, we can discover the shortcomings in our real self. Our self-inflated ego is the best sign of our immaturity.

The more we inflate our ego, the more we increase our anxieties and fears. One of the reactions to any increase in fear is to escape, it is this tendency which transforms those with inflated egos into restless and agitated people. We are, in fact, the most restless and agitated species on the planet. We spend a great deal of our resources on tranquillisers. This consumption of tranquillisers increases in direct proportion to our mental pretension.

Because of its anxiety, man's inflated ego becomes strongly self-centred which prevents him from listening to others. Lacking the quality to listen, this self-centredness hinders an understanding of others, thus preventing tolerance, co-operation, coexistence, civility and compassion. Borne along by the inflated ego, and particularly in the 'macho' cultures, the obsession with sexual conquest and prowess increases. Self-righteousness increases with an inflated ego and with this increase in self-righteousness it is inevitable that sexual violence and rape become more prevalent.

With the appearance of the mind and its pretensions, two very curious phenomena unique to man appeared: prayer and beggary. We pray and beg the supernatural divinities, creations of our mind, to gratify our pretensions, to a degree far beyond our just deserts. Crying, which is typical of infancy and an infantile mentality, is practised mainly by people when their minds' wishful desires or pretensions become frustrated, shaken, or when they collapse.

What is wrong with living in the supernatural world of the mind, the world of fantasy and fairy-tales?

The main fault is that the infantile world of the mind perpetuates infancy, its selfishness, self-centredness, aggression, violence, destructiveness, cruelty and amorality. The innocence of infancy or the infantile mentality is another wishful belief of the mind.

Another negative side of the mind is that it creates mental disorders and psychosomatic disease.

The fault is also that the supernatural tendency created by the mind runs contrary to nature and its harmony. The best evidence of this is the increasing endangerment of life on our planet by man's wishful beliefs which started when God entitled man, still in his infancy, to rule and to dominate the planet.

Another negative aspect of the mind is that it creates an agitation and restlessness which run contrary to wisdom and serenity, the necessary condition for a healthy well-being.

In fact, our mind can seriously damage our health. Day-dreaming and dreaming, particularly the dreaming during sleep known as Rapid Eye Movement or REM, have similar effects on our body. These effects are: an increased activity of our sympathetic nervous system, an increase in heartbeat rate and blood-pressure, more rapid respiration, higher arousal and tension, increase in the consumption of oxygen by our brain, and an inhibition of the gastrointestinal activities. Our day-dreams as well as our dreams can be incongruous, and deceptive. Our belief in miracles must have been inspired at least in part by miraculous dreams. We know how damaging our dreams of miracles can be. In fact, these beliefs in miracles inspire our credulity and our passion for the ideologies promising miracles. Belief in miracles can also increase our adventurousness which can, in turn, place us in situations in which we need real miracles to save us.

One of the most tragi-comic inventions of the mind is the cult of individual freedom and the cult of individual independence. Millions of people died fighting to realise these two illusions.

Any rational brain would realise that individual freedom and independence are not in the interest of our social species, as they can be realised only at the expense of other members of the same species, of our family, of our 'domus', or our community. Individual freedom and independence are abstract ideas created by our mind in order to help itself to fantasise and to day-dream more freely and more independently, to perpetuate itself and to achieve greater power. There is no such thing as individual freedom and independence in nature: nature is ruled solely by the laws of nature. That is why the mind's individual freedom and independence cannot do anything other than fly in the face of nature.

With freedom and independence man tries to escape from reality into his surreal mind's world of self-adulation and self-deception. Man tries to escape the real world because he fears that his shortcomings, his fragility, will make him appear ridiculous in the presence of nature.

What is ironic is that man falls in love with the mind's wishful ideas, beliefs and ideologies. In this way, he becomes the slave of these same ideas and beliefs, thus losing his freedom and independence. A strong believer or a fanatic is not a free or independent man. Man pays a heavy price for his escape into the illusion of individual freedom and independence. These illusions create a loneliness which in turn increases man's fears and anxieties. It is known that any increase in our fears and anxieties can damage efficiency of our immune and repair systems. There is positive evidence that sick and wounded people recover more quickly and efficiently from disease or injury if they are convalescing with their families rather than in isolation created by a desire for individual freedom and independence.

In the name of his individual freedom and independence, man runs contrary to the interest of the community, he places himself in a position of competition, often ruthless antagonism or rivalry against the other members of his own community which can only damage the community's co-operation, damaging to everyone's best interests. In order to justify itself, our mind invented the greatest fallacy: the absurd belief that ruthless competition is a more efficient *modus operandi* than a civilised and intelligent co-operation among the members of a community. In fact, the very essence of a social species is the co-operation amongst members of that species.

Co-operation inside a colony or a group of a social species creates the colony's intelligence. This intelligence is mainly concentrated on regulating and organising the colony's individual members' activities in the interest of the community, which interest is in turn in the best interest of the members of the colony. In fact, organisation is the brain of the superorganism which constitutes the colony or community. Like any other brain, the brain of a superorganism tends to evolve towards better ways of long-term survival. A cell's intelligence consists of its organisation of the activity of its components. A brain's intelligence consists in the organisation of the co-operative activity of the neurons. If an organelle of a cell or a neuron of the brain decided to develop individual freedom and independence and thus selfishness, this would bring an end to the cell and thus the end of the brain's rational capacity. If a cell of an organism develops selfish individual desire for freedom and independence, like the cancer cell, that would spell the end of the organism. With his selfish mind, man seems to be the cancerous cell of the superorganism which is life on our planet.

We can compare a social species' community's superorganism and its intelligence with a symphony orchestra. If the members of the orchestra were free and independent agents, allowed to play their instruments in a selfish way and purely at random, they would produce a cacophony. Only through the organised co-operation of the musicians can we have harmonious music.

In our obsession with individual freedom and independence, we have created, and we continue to create, an ever-increasingly chaotic and disordered world. But in its infantile mentality, man's mind is both excited and attracted by chaos and disorder. Order bores infants. Some people even believe that their God will start redeeming them only when they have sunk to the lowest levels of decline, and this belief can only lead to those lowest levels.

In his obsession with individual freedom and independence, man has placed himself in a position of solitary confinement in which fear dominates, which is the real hell of which Sartre speaks.

In this fear, man develops a here-and-now attitude and behaviour, an instant life, he concentrates on the omnipotent and the omnipresent "now" focusing on immediacy. This life tends to wipe out our past and our future, it tends to annihilate time, and without time there is no real existence. Without time there is no sense of lasting, no patience, no pondering, no reasoning, no serenity, no wisdom, no sacrifices and no generosity. To live with an awareness only of the present also confounds the education of future generations. Living in the present transforms the global economy into an activity concentrated more on quick exploitation and an instant profit rather than on investment. A life in the here-and-now, a life with no sense of the future, prefers consumerism to saving or sacrifice. Instant life is dominated by instant excitement. Instant life, created by the illusion of individual freedom and independence, has created an 'instant' family, an 'instant' relationship, 'instant' emotions and even 'instant' pleasure.

Through ignoring the past and the future, we are more and more deracinated, and, like a tree, a deracinated person cannot bear fruit. Without roots man is suspended in the emptiness of his mind.

This instant life, concentrated on the selfish and ruthless present, contributes more and more to the frustration of the old population, a population without a future, with a past constructed of a chain of selfish and ruthless presents. This sorry past is seen nostalgically because we glorify it through fugues of mental fantasy. The mind is a greatly skilled artist in self-deception.

Isolated in this solitary confinement, selfish and ruthless and independent individuals invented their perfect ideology, capitalism, which is based on individual selfishness, self-centredness and ruthlessness. (In a later chapter we shall examine in detail the concepts of profit and profiteering.) Karl Marx was wrong when he insisted that it was capitalism which created alienated individuals. It was, instead, self-alienated individuals that created capitalism. Capitalism continues to alienate people simply because it was created by alienated people. The supreme aspiration of any ideology is to perpetuate itself, to father its wishful beliefs.

These short-sighted individuals, living in their fearful present in which rationality can seldom come into prominence, were the fathers of the Industrial Revolution, so glorified in human thinking, but which, like most revolutions, was more damaging than beneficial. Among the negative aspects of the Industrial Revolution we can cite the following: an increase in stress and stress-related diseases and mental disorders; a proliferation of depression and suicide; the appearance of slums and their attendant poverty; more prostitution and consequent venereal disease; a general increase in crime and violence, brutality and vulgarity, alcoholism and amorality. Industrial Revolutions transformed workers into the slaves of machines. Our most recent

Industrial Revolution, still in progress after three hundred years and showing no sign of abatement, brought with it the most damaging pollution, in particular global warming, which sooner rather than later will surely prove to be catastrophic.

The Industrial Revolution drugged our minds with technological gadgets and toys and speed; gadgets to play with in our solitary confinement and speed to try to escape pondering on the fact that our pretentious mind has brought us, and other species, to the edge of the precipice.

Another negative side of industrial revolutions is that they obliterate or deform our sense of the past. Without a sense of the past, or with a deformed memory of it, the mind is dangerously freer to give its fantasies full rein.

In fact, our mind tends to manipulate our memory and to select from it and from the past only that which reinforces its beliefs, ideologies or prejudices. That is, perhaps, why history cannot teach humanity how to avoid repeating the same mistakes again and again.

The selective nature of the mind as far as it concerns the memory can best be seen in the fact that people who change their beliefs or ideologies try to find in their memories or in their historic past facts and events which fit the new belief or the new ideology they wish to adopt. They forget the facts and events which served the previous belief, the previous ideology. A believer's forgetting is a voluntary act of the mind. After all, it is the mind which invented mental agility and it became proud of it.

Our mind can also prevent our central nervous system from perceiving facts and events from the outside world, thus avoiding the need to memorise them, if they can influence a belief, ideology or prejudice: such is the power of the mind. The mind is powerful because its potency is based on willpower, inspired by the wishfulness of the fragile and frightened creature who is a believer. Nietzsche's Superman is, in essence, a poor immature, undeveloped, frightened, stubbornly pretentious believer.

Man pays for his clever escape from reality into the mind's world of wishful beliefs by becoming fearful of his future. A person with a selective, invented or manipulated past starts doubting his future; this goads him on in his ruthless exploitation of the present.

Living in the mind's strong belief-system limits our curiosity, exploration and learning which tends to perpetuate an infantile mentality, which prevents mental and cultural maturity.

Living in the mind's world of beliefs created an ugly hatred of anything or anybody which might oppose or contradict our beliefs. The atrocities of inquisitions and persecutions have always been perpetrated by strong believers. The stronger the mind, the more inclined it is to develop hatred. Man is the only species with the ability to hate. Hatred as exemplified by man is not to be confused with traits of survival aggression in other species, the stalking lion or the venomously biting snake. Man's hatred is a mental fabrication, superfluous to any constructive natural scheme.

We notice that in any community those who are more fragile, more vulnerable, more immature, more helpless, live more in the clouds of their minds than those who are the opposite. As they are more mature than men, women are less seduced by fantasising and by philosophising. Women regard men's propensity to philosophise as a childish game. That is why philosophers hate women and their common sense.

Even whole communities or nations are liable, in periods of instability or uncertainly, to intensify the desire to fantasize. Lenin, Mussolini and Hitler built their powers on the day-dreams of their people, exploiting the chaotic situation in their respective countries after the First World War. It was the misery of the Russian people at the end of that war that spawned popular credulity for the miracles that Soviet leaders promised to perform. Prophets prosper in times of hopelessness.

With the development of the mind we acquired another trait unique to our species: stress-addiction. In fact, many people thrive on stress. This is, perhaps, due to the release in moments of stress of endorphins, heroin-like molecules, in our brain, which produce a pleasurable sensation. Many people also develop stress-addiction because, in its frailty, our inflated ego craves a sense of importance and we feel important when we are over-scheduled, when we are agitated, when we are rushing, states which are stressful and fear-inducing.

If we accept the idea, which I stressed before, that life has been imposed or forced on some inorganic elements by the energy of the sun, transforming these elements from the more stable existence of inorganic matter to a more unstable, a more irritated and agitated existence in an organic state, than it would be logical that organic matter must carry within itself two main tendencies: first, the tendency to return to the previous state of reduced instability, lower agitation, and second, the tendency to prevent or to avoid any further instability, irritation or agitation.

If we accept this idea, then it follows that the prime organisation an organism has to develop it is its defence mechanism aimed at reducing the existing instability and the prevention of any further agitation. The defence mechanism of any living organism lies in its immune and repair systems. These systems are essential and their collapse would result in the collapse of the organism itself.

If this is valid, it would then be logical to suppose that it is our defence systems which influence our organism's growth and development, and the end of them. It implies, then, that it is our defence mechanism which influences the development and the organisation of our senses and of our brain. In fact, when our brain is in a state of deep sleep, and thus cut off from the real world, it is our defence mechanism which dictates our survival.

In the rest of the animal world the brain's activity is in tune with the activities of the immune and repair systems of the creature's organisms. In a mature organism the brain and the defence mechanism have the same logic, the same objective, which is to defend and to protect the organism.

But, because it is dominated by the irrational mind, man's brain can easily damage his immune and repair systems. Imposing itself and its irrationalities on the brain, and forcing the brain to serve or to obey the mind's inflated, capricious infantile ego, our brain runs the risk of not being in tune with the objectives of the defence mechanism. By serving the mind's interests, our brain, can, in fact, damage our health by causing psychosomatic diseases, cancers, heart problems and auto-immune diseases or disorders.

Our mind has a damaging effect on our immune and repair systems, chiefly in the following ways:

Living in the mind's world of illusions, wishful beliefs and day-dreams creates a certain anxiety which in turn creates a state of emergency. This triggers off the activity of the hypothalamus which activates the emergency endocrine system and the sympathetic nervous system, placing our body under stress, and this inhibits the efficiency of our immune and repair systems, rendering our body susceptible to the aforementioned diseases. In fact, we are open to more diseases and disorders than any other species.

Other animals behave in the same way when faced with an emergency. The difference is that they try to fight, to flee or to hide from danger, thus reducing the duration of the emergency to the briefest period. Our states of emergency tend to be prolonged for dangerously extended periods, often the major part of man's life, as we seldom fight, flee or hide from our *mind*, the primary source of the emergency. What is more, we are even inclined to court danger, since the mind enjoys excitement provoked by adventurousness or risk-taking. We reward our soldiers with medals and other honours for placing themselves in the line of fire, and in other walks of life we applaud those who are able to withstand prolonged stress, upholding them as admirable examples of how life should be conducted.

A strong, rigid and inflexible mind also inhibits the efficiency of our senses, which in turn limits our brain's efficiency in perceiving external signs of danger. In fact, we are the most accident-prone species. The stronger and more inflexible the mind, the more accident-prone it is.

I have stressed before that for millions of years our human ancestors' communities were guided by mothers and particularly by the more experienced elderly mothers. I have also stressed that the mothers' main attributes are: loving, sharing, nurturing, nursing, sacrificing, protecting infants, and taking care of the sick and injured. Since our male ancestors were trapped in infancy, we have only survived as a species because in our ancient communities these maternal attributes came to the fore. Mothers also have an innate sense of organisation and planning.

That our ancestral mother's brain was better equipped for the survival of our species can be deduced by looking at the present difference between the female and the male brain. A woman's brain seems to be more mature and better organised than that of a man. A woman's brain is able to cope with several activities simultaneously while a man tends to concentrate on one thing at a time, often to the point of excessive concentration and distraction. This gives women a wider, deeper and a more comprehensive judgmental and reasoning faculty, and therefore a more accurate notion of the real world, a better vision of the future and a more realistic appraisal of events, scenarios and sensations. It enables her to reach a more profound understanding of others and to develop the key instrument of compassion so vital in a species that more than any other is prone to the procreation of unfit, ill-equipped individuals.

Women have more efficient senses, a better perception, a wider memory capacity and more accurate intuitive faculties than men. This enables them to enjoy a richer participation in life. They have better linguistic and communicative capacities, which is important in a social species that habitually provokes aggression, violence and war because of poor understanding or bad communications. Women's brains have more highly-developed centres dealing with loving, caring and co-operation than men's, vital in a species producing an ever-larger number of senior citizens.

A woman's brain has another important component, which is becoming more and more vital for the survival of our planet: she has an innate sense of cleanliness, tidiness and purity.

She has a far greater predisposition for work than man. That is perhaps why she is likely to be more generous, more inclined to share.

A better plasticity in the connections and the networks of her brain cells affords her a superior perception of environmental changes.

She is nearer to nature, and therefore more cosmically aware, more universal and more cosmopolitan than a man is. A man tends to hide himself and his shortcomings behind his provincial prejudices or the wall he has built with the bricks and mortar of Chauvinism.

What, then, took place historically that created a new situation in which the less developed and less mature male succeeded in subjugating the more developed and more mature female?

Some men who developed strong minds entered into a phase which could be called the mind's adolescence, a phase in which men's minds started taking their beliefs, fantasies and imagination over-seriously. Men placed a blind faith in their beliefs which transformed many of them into fanatics; aggressive, violent and cruel militants.

In this state of adolescence, man created in his fantasy an idealised self; his ego. In his mind he saw himself as superior to woman. The precariousness of his belief in his superiority, and the fear which is inherent in any wishful belief, imbued man with an extra nervous tension, and an extra energy which he channelled into the materialisation of his wishful belief. In fact, the stronger a man's mind and his beliefs, the more violent, aggressive and cruel he becomes.

That man's idealised self is in a permanent state of emergency can be seen by the fact that it tends to display itself, to show off. In nature, animals display their feathers or inflate their bodies when they need to impress or to intimidate those who threaten them.

Some of these male fanatics become leaders of their mentally infantile fellows. These infantile followers gain confidence in their wishful beliefs when they are in tune with those of the strong-minded leaders who have had the audacity to place a blind faith in their beliefs. In fact, it is often the case that the more fragile and infantile men are, the more fervently they will follow a fanatical leader.

These fanatical adolescents, who started what can be called an adolescent revolution, succeeded through their aggression and violence in imposing their patriarchy on human communities.

The supreme aim of this patriarchy was to subjugate women and prevent them from deflating male mental pretension. Men debased the value of women's attributes, her predisposition for domestic work and so on, which were the essential factors in the survival of our species.

She became his major enemy, an enemy to be conquered, and even eliminated if necessary. In an Ancient Sumerian myth, we see Marduk, the first important male God, killing his own mother, in order to seize her supreme power.

In his mind, man invented the idea of the dominant father. He also invented the idea of the family dominated by the omnipotent father. Man had to invent the patriarchal family because it did not previously exist in our ancestors' communities. Since it was based on man's wishful ideas, patriarchy had to be implemented by force. Punishment was meted out to those disobeying the new order. A woman's adultery or disobedience might even be punished by death. Wife-battering and the abuse of women in general was manifested the world over in a variety of carefully legislated practices, from the stoning of adulteresses in the Muslim world to the persecution of witches – and who were they but custodians of primeval maternal wisdom? – in the west.

For millions of years our ancestors' communities were dictated by imitation of others' example. Having no blue-print to emulate, man set about creating laws and commandments which had to be obeyed. Teaching became based more on preaching than on examples to be imitated. Man became a slave of sterile and overweening precept rather than a constructive modifier of practice.

In the first important stage of development of man's infantile mind, in one of man's favourite fairy-tales, the Bible, man invented an omnipotent male God who created man in his own image and who then made woman in order to please and to serve man. The Judaeo-Christian belief-system even demonised woman. It also trivialised woman's virtues and her domestic work.

The Ancient Greeks, who glorified the mind and had a great propensity for philosophising, denigrated women to the lowest level. The negative attitude of Plato and Aristotle, the two philosophers who shaped Western culture, is well known.

The fathers of the Christian Church were extreme misogynists. Clement of Alexandria insisted that 'all women should die from shame at the mere thought of being a woman.' St. Augustine explained that 'If it was good company and conversation that Adam needed, it would have been much better arranged to have two men together as friends, not a man and a woman.' Even Luther insisted that 'women were created only to serve and help men.' Darwin stressed that 'man has ultimately become superior to woman,' and that woman, this 'undeveloped man' is a serious obstacle in man's evolution towards superior forms of his predatory nature. Darwin belonged to a culture dominated by men who in turn dominated the world: in the culture in which he flourished, mothers were in thrall to the males of the family, wives were subservient to their husbands and daughters worshipped their fathers.

Nineteenth-century male-dominated culture invoked the arts as a means of debasing women. Nude paintings of women became fashionable. In the paintings of this period we usually see a woman sleeping, swooning, convalescing, mad or dead, all states in which she can pose no real threat. At best she is represented as some form of romantic ideal, carefully shaped in order to fit in with the dominant scheme that man has tabulated for himself.

In order to be able to escape fully from reality into the world of fantasies and to see themselves in their imagination as greater and more powerful than they were in reality, men needed free time, time to day-dream. Paradoxically, it was women who provided men with this free time, in which they were able to discover a life of fantasies, and to play games inspired by an infantile imagery. This must have developed on a larger scale with the Agricultural Revolution, which started twelve thousand years ago. This revolution must owe its genesis to the domestic stability provided by women. Women were more home-orientated, and therefore started settling in one place: nesting is in a mother's nature.

Once she settled, she began cultivating the land around her settlement and domesticating animals. In fact, the first agricultural divinities in history were women.

Our ancestral mothers chose the creatures which were to become our domestic animals, creatures that were more immature, confused in their instincts, easier to domesticate than more highly-developed species (highly-developed in the sense of possessing a focused mechanism for survival that prevents them from being susceptible to domestication).

Our ancestral mothers' natural predisposition to nesting, nursing, protecting, caring, sharing, economising and storing food, helped them to achieve this Agricultural Revolution, which reduced man's need to spend most of his time searching for and gathering food. This Agricultural Revolution helped men to realise their Adolescent Revolution, in which their adolescent arrogant mentality replaced women's maturity and their common sense reasoning.

Having more responsibility and being more occupied with the realistic problems of life than a man, an adult woman seldom has free time to indulge in the world of fantasy, in the world of wishful beliefs.

Not having developed wishful beliefs and fantasies on a scale in which they would see themselves more important than they were in reality, women seldom develop a hatred of men. Those women who are able to find free time in which to fantasise are also able to develop pretensions, conceits, infatuations and hatred.

In his search for the consolidation of his power and for the total subordination of woman, man found a great ally in his religious beliefs and in his monotheism.

In his hatred of woman, man invented his monotheistic God without a mother. This is contrary to the basic law of nature.

Judaeo-Christian monotheism is, in essence, a creation inspired by man's fear of woman. Denigrating woman, reducing her to obedience and reverence for her man, and thus to man's invented God who created man in His own image, is what monotheism is all about.

Monotheism introduced and consolidated concepts such as absolutism, autocracy, patriarchy, hegemony, tyranny, ruthlessness, intolerance, hatred, rigidity, stubbornness, 'others' and 'otherness', fanaticism, 'who is not with me is against me', supremacy, superiority, hierarchical relationships, the consideration of work as a punishment and the consideration of personal sacrifice as a humiliation. All these concepts are created by man's mind.

This monotheism helped man to replace the natural superiority of woman with his religious belief in his own superiority, The strength of a wishful religious belief rests in the fact that it justifies any means, mainly brutality, violence and fanaticism (so feared by woman), to realise its goals. Man's God considers 'holy' or 'just' any means or any use of force, brutality or violence against those who oppose the aspirations and interests of His male believers. The witch-hunts of the past are the best evidence committed again women by men in the name of God.

Man's subordination of woman is a clear sign of the inferiority of his mentality. The supreme aim of an inferior mentality is to achieve superiority. An inferior mentality seldom aims at equality. In the eyes of an inferior man, equality would leave him with the feeling of inferiority.

A man with a pretentious mind also tends to despise and to hate a woman because he needs her love, care, protection, advice, company, her very existence. In fact, God created Eve because He felt that being lost, lonely and not self-sufficient, Adam needed her help. In his infancy, Adam was hopeless.

Since his successful Adolescent Revolution, man started considering his woman as his possession. Like any other possession, his woman could be exploited, used, abused, beaten and eliminated when not needed any more. In order to reduce woman's identity, to transform her into his property, man imposed on her his own name, the ultimate sign that she belonged to him. Woman became forced to please her man sexually and to gratify his increased sexual obsession. In order to please her man, she even started to pretend to enjoy sex, a pretence that is not evident in any other species. As the fantasies and imagery in his mind became more shaped and more focussed, man's obsession with eroticism increased.

After his successful Adolescent Revolution man started to despise work even more. He felt that work was debasing his inflated ego. What is more, work was limiting the mind's fantasising as it brought man face-to-face with reality. (In fact, as work can reduce fantasising, it can be good therapy for some mental disorders). In the Bible (this big man's mind creation) work is considered a punishment.

It is interesting to notice that cultures which most degraded women also despised work. In Ancient Greece, where the human mind reached high levels of philosophising and day-dreaming, women and work were highly denigrated. In Plato's *Republic* we read: 'It is fitting for man to despise work'. To Aristotle 'all manual works are without nobility; it is impossible to cultivate virtue and to live as a wage earner.' According to Xenophon, Socrates said: 'The workers and their handicrafts are despised and discredited in the cities."

In Ancient Rome, women and work were more respected. In Rome, woman was a real 'domina' in her 'domus'. She was in charge of the family life and the guardian of domestic gods. King Numa Pompilius used 'to judge people by their work'. Plautus suggests in *Persea*: 'Age quod agis', meaning, whatever you do do it well. Seneca stressed in his *Epistulae Morales*: 'Nihil est quod non expugnet pertinax opera, et intenta ac diligens cure', meaning that there is nothing that cannot be won with persistent work, and attentive and diligent care. In his *Georgics*, Virgil insisted that 'Labor omni vincit improbus', meaning that hard work beats everything. All of these classical perceptions are embodied in what in our own culture we know as the Protestant Work Ethic, but essentially the informing idea behind our interpretation of both Greek and Roman precedent is the wilful subjugation of women.

Aggression

Every living organism has an innate drive to defend itself. This defensive predisposition is the part of an organism's tendency to fight or to try to avoid any new biological discomfort or injury.

With the development of the mind, man acquired a new phenomenon in nature: his offensive aggression. This offensive aggression is destructive because it consists of the mind's surreal belief trying to become reality. This can happen only by destroying or disorganising reality.

As man's mind is able to be devious, vicious, brutal and ruthless, his offensive aggression tends to follow the same line.

Having a better organised brain, thus having a less fertile mind, a woman is far less offensively aggressive than man. Women who try to imitate men can often be as violent and as ruthless in their aggression as men.

There are two main views regarding the origins of man's aggression. To Instinctivists, our aggression is phylogenetically programmed, an instinct. For this school of thought, life is a jungle in which the fittest, whatever that may mean, has a better chance of survival and of transmitting his aggressive genes to future generations. To Behaviourists, human aggression is mainly caused by social and cultural environments.

Some representatives of the instinct theory claim that our aggression is highly beneficial. In his book *On Aggression*, Konrad Lorenz wrote: 'Summing up what has been said in this chapter, we find that aggression, far from being the diabolic destructive principle that classical psychoanalysis makes it out to be, is really an essential part of life, preserving the organisation of instincts.'

In Spengler's *Man and Techniques*, we read the following: 'The beast of prey is the highest form of active life... The human race ranks highly because it belongs to the class of the beast of prey' ... 'the life of a man is the life of a brave and splendid, cruel and cunning beast of prey. He lives by catching, killing and consuming. Since he exists he must be master.'

In Bertrand Russell's *Authority and the Individual*, we read: "The old instincts that have come down to us from our tribal ancestors, all kinds of aggressive impulses inherited from generations of savages'...

I would like to point out to those who insist that man was and is a born 'hunter and killer', that nature would never have given the predatory instinct to a species with the digestive system of a vegetarian, like other primates. If man was a 'beast of prey' he would have never have turned his hand to agriculture, the domestication of animals and the dairy industry.

The following facts could help in proving that the origin of our destructive aggression lies in our minds.

Ablation of the frontal lobe where the wishful way of thinking seems to take place reduces offensive aggression.

Offensive aggression can be manipulated by influencing the mind with propaganda or brainwashing.

In order to arouse aggression in their followers, political and military leaders depict their opponents as monsters, criminals or dangerous war-mongers. Conflicts are sustained by the energy generated by a fear of defeat, and a fear of defeat by 'monsters' increases emotional arousal and aggression even more.

Hysterical delirium of the masses, created by the manipulation of their minds, can lead to the most atrocious destruction and savagery, a phenomenon explored early on enough by Euripedes in *The Bacchae*.

Superstitions, ideologies and religions are among the main sources of offensive aggression.

When the mind becomes obsessed by a belief or an idea, it develops fanaticism, a serious cause of destruction.

Tyranny, the epitome of offensive aggression, is mainly based on an ideology.

The worst types of offensive aggression are manifested in persecutions or revolutions, and these are inspired by an idea or a belief.

Other major instigators of crime and violence are racial or national prejudices, all creations of the mind. People inspired by these prejudices often justify their killings with the belief that they are not eliminating members of their own species, other human beings, but 'barbarians' or '*Untermenschen*'.

Jealousy, envy, vanity, resentment, malice, hatred and spitefulness are all sources of offensive aggression, and are all states of mind, or emotions created by states of mind.

Throughout time and space, the torture of man by man has been committed in the name of a belief or a prejudice. Inquisitors are always ardent believers.

Man considers a moral insult a good reason for a strong reaction. But insults stem from excessive self-esteem.

Honour and manliness are the creations of the mind and they are an important source of offensive aggression.

Some people insist that frustration is at the root of most of our offensive aggression. Frustration, however, is nothing but self-infatuation offended by reality. We are frustrated in a bus or train full of people we consider to be ugly and ordinary. We are frustrated by working in a crowded office or factory. We are not frustrated, however, by a much bigger gathering at a reception at Buckingham Palace or the White House, or a fashionable night club. A man is seldom frustrated by a crowd in which he feels important, which is applauding him, carrying him on their shoulders or pressing him for his autograph.

Self-confidence is another significant source of destructive aggression. It finds its energy in the gap between individual pretentiousness and feelings of inadequacy.

The glorification of self-confidence invaded Western Europe with the Renaissance. 'It is better to be adventurous than cautious', wrote Machiavelli, 'because fortune is like a woman, and if you wish to keep her under, it is necessary to beat and ill-use her.' Machievelli certainly did not know that self-confidence in a life ruled by uncertainty and unpredictability ends in a stubbornness derided by ironies. Corneille was much wiser when, in his *Cid*, he explained that 'Danger breeds best on too much confidence.'

In the last two centuries, the Western World seems to have been in a phase of what could be termed suicidal aggression. This suicidal aggression is perpetuated in the name of another wishful belief of the pretentious mind which is pompously called 'progress'.

The degradation of the quality of our life and the pollution of the planet and its atmosphere, caused by aggressive progress, is proving more and more that the mind's illusions can end in dangerous and suicidal delusions.

The World Health Organization foresees a serious increase in depression in the Western world throughout the course of the next century.

The Arts

Within the development of the mind, man developed what we call the fine arts. These fine arts are mainly created through fantasies, imagination and fictionalisation.

Being based on the mind's reveries, the arts' main contribution consists of the stimulation and enrichment of day-dreaming, the escape from the real world. Visiting an art gallery can bring certain people to the outer boundaries of the real world, causing them to faint or to feel dizzy. This is known as the Stendhal Syndrome, since it was the French writer who was the first to describe the phenomenon.

Lovers of art insist that art has an ennobling value. This cannot explain, however, why so many art lovers or artistically creative people were brutal criminals. Many Nazi leaders were passionate music-lovers, and some of them were great art collectors. Many artists have been arrogant and aggressive. Periods of the most atrocious crime and violence, such as the Fifth Century BC in Greece, and the Fifteenth and Sixteenth Centuries in the Western World, coincided with a flourishing of artistic creation. The Spanish Inquisition coincided with the golden age of Spanish art. The Nineteenth Century and the first half of the Twentieth Century were rich in artistic creativity, but also produced a series of colonial wars, the two major world wars and countless other persecutions, exterminations, concentration camps and gulags of all kinds.

As she is less attracted by the mind's fantastical properties than a man is, a woman tends to be less artistically minded. The Ancient Greeks felt that offensive aggression and the arts were related when they named the goddess Athena as the protector of both. They also knew that aggression and the arts were products of the mind when they claimed that Athena sprang fully-armed from Zeus's head. 'At the stroke of the bronze-heeled axe, Athena sprang from the height of her father's head with a strong cry. The sky shivered before her and earth our mother too.' It is interesting to note that in his *Eumenides*, Aeschylus attributed the following proud statement to Athena: 'No mother bore me, in all things my heart turns to the male, save only for wedlock, and I incline wholly to the father."

Athena, the supreme goddess of men's offensive aggression, and of their fine arts, was, therefore, the daughter of a male. She had no mother. Later scholars have not failed to identify the aggressive tension that is evident in the development of Greek culture, and they refer to it with predictable admiration as the 'agonal' element, the implication being that aggression, struggle, is a component part of the production of something worthwhile, whether it be simply a work of art or more ambitiously an entire social structure: men, not women, should customarily be its agents. Athena's abnegation of the laudable female characteristics we have listed (except crucially that of entering into wedlock) is but one of the many examples of the ways in which mysoginistic thinking has been refined and perpetuated by the male of the species over the millennia.

Happiness, Excitement and Boredom

In the animal world we observe the pursuit of a lower degree discomfort, of reduced discontent. Using our brain and its intelligence, we can acquire wisdom by limiting our pretensions inside the limits of our abilities to realise them. Inspired by wishfulness, wants and desires and ruled by pretentiousness, we seldom reach contentment because the needs of a pretentious mind are unlimited. Pretensions tend also to increase with the gratification of the mind's desires. Driven by pretensions, our mind tends to develop a more or less permanent sense of discontent.

Could it not be that being infantile and thus susceptible to growth, our mind's pretensions tend to expand permanently? In this expansion, our mind does not allow our brain to develop humility and gratitude for what we already have. Humility and gratitude can diminish discontent. Everyone can realise humility and gratitude just as everyone can introduce into his brain the rational and realistic concept that things could always be worse. The mind's idea that it could always be better perpetuates discontent.

Pretensions are also dishonest. They are aspirations that cannot be realised except at the expense of others. But, the mind's world is an amoral world.

At a certain stage of its evolution, the mind invented the idea of happiness. A wishful and therefore aggressive idea, the idea of happiness gave rise to the *pursuit* of happiness. The pursuit of happiness is, in essence, a pursuit of the mind's pretensions which is futile for the following two reasons. First, the mind's pretensions are beyond our abilities to realise them. Second, when, by some miracle, we succeed to realise our pretensions, we become even more pretentious, even more unhappy.

The pursuit of happiness develops restlessness, agitation and over-activity. As it is closely akin to the state of emergency, over-activity prevents meditation and rational endeavour, things which are in the mind's interest. That is, perhaps, why over-activity can become an addiction and an obsession. That over-activity can limit the efficiency of our senses and of our intelligence can best be seen when it ends in panic. Also, the negative side of over-activity and panic is that they are both very contagious.

In their hyperactivity the Americans failed to see the irony of naming the pursuit of happiness as one of the 'rights' in their Constitution. It seems to make them consider themselves entitled to and justified in their obsessive over-activity at any rate.

Many people transform the pursuit of happiness into the pursuit of excitement, excitement consisting of arousal caused by the increase in the activities of our sympathetic nervous system and the release in our body of some of the emergency hormones and neurotransmitters. Some of these neurotransmitters help the release in our brain of our natural opiates which can provide pleasant feelings and even euphoria.

We tend to assist in sporting events not so much as to enjoy the beauty of the games but to merely get excited. We get worked up by becoming fans of one or other of the players in the game. Becoming involved in a game as a fan, we develop fear or worry that our side might lose. It is this fear or worry which stimulates our excitement. During this excitement, we often vent our arousal, our nervous energy (accumulated in our body by fear or worry of losing) by shouting and screaming and often by aggressive or violent behaviour. That is why excitement can also be dangerous and barbaric.

It is in the nature of the mind to love excitement because, limiting the efficiency of our senses and the reasoning activity of the brain, excitement increases the mind's fantasising and day-dreaming.

Obsession with excitement can transform us into becoming attracted by disasters or catastrophes. This attraction is unique to our species. Other animals are frightened by disasters and they try to escape them.

Being fascinated by disasters we incline to cause them. Some religions go so far as to promise that God will redeem His believers when they have sunk to the lowest levels.

When remembering events, we tend to better recall defeats and tragedies than happy occasions. With regard to the Bible, we tend to remember the Great Flood. The name of Jesus Christ reminds us more of his tragic death than of his resurrection. Likewise, Caesar's name is associated more with his brutal end than his glorious conquest of Gaul. When recalling Dante's *Divine Comedy* we remember more the chapter on Hell than that on Paradise. The Nazi extermination camps have more visitors than many of Germany's important cultural centres.

Perhaps, we like disasters and tragedies happening to other people because faced with them we feel less discontented within ourselves. The failures of others seem to be a solace to us.

Many people find a great excitement in watching the execution of those condemned to capital punishment. Until the last Century, many white Americans found excitement in assisting in the lynching of blacks.

Participating with others in great tragedies or disasters, we often develop a feeling of joy or merriment. Perhaps, being by nature a gregarious species, we feel, somehow, more protected in sharing a disaster.

Having a more developed mind than woman, man is more obsessed with excitement than woman: and particularly as far as the most irrational of his excitements are concerned, which are his adventurousness, his risk-taking or his gambling.

It is interesting to note that a great deal of our civilisation has been created as a result of irrational adventures. Many people might doubt this assertion, mainly because our culture, which is dominated by our minds and their interests, considers the spirit of adventure as one of the most glorious of man's virtues, the supreme attribute of courageous manliness.

In his present infantile mentality, man is also attracted by cruelty, by macabre, grotesque and bloody scenes. In her maturity, woman tends to find these spectacles both frightening and pathetic.

People with a passion for excitement long to have more and more free time in order to be able to become more and more exhilarated. Ironically, we tend to call this free time 'leisure time', ironically, because, during leisure time, instead of trying to achieve serenity, many of us actually step up our pursuit of excitement. It is ironic too, that in this leisure time we see an increase in brutality, aggression, violence, rape, crime, destruction, depravity, hooliganism, perversion, gambling, alcohol and drug consumption and vertiginous driving.

Modern industry concentrates more and more of its activity on providing more and more efficient gadgets able to provide extended leisure time, in which we can become yet more restless, agitated, stressed or depressed.

With the development of the mind, man acquired the enjoyment of being entertained. Amusing, bemusing or engrossing, entertainment helps the mind increase its capacity for reverie and fantasy, to escape from the 'sad reality' which is offensive to the mind's inflated ego. In truth, our ego feels important when it is being entertained.

Our obsession with amusement and excitement inevitably creates boredom. When we are bored we crave yet more amusement and excitement. Boredom engenders irritation and aggression. The mind is irritated by boredom because boredom offends the mind, it undermines the importance of its ego.

The more inflated and infantile a mind is, the more easily it becomes bored. It is also more boring. That is why bore and bored are often the same person.

Men are more susceptible to boredom than women. Those women who imitate men can be even more susceptible to boredom than men. (They can be also extremely boring!)

Altruism

Many people have tried to explain the development of altruism in our species. Some forms of unselfishness such as sharing food, the protection of the community, or taking care of infants, are also present in some other species, but our concept of altruism goes over and above anything that is associated with the natural order of things.

Before the pre-eminence of the mind, our ancestors' communities were guided for millions of years by mothers. In this kind of matriarchy, the mothers' behaviour, which tended to be altruistic, was imitated by the rest of the members of the community. With the Adolescent Revolution, and with the development of cults of individual freedom and independence, individual selfishness became increasingly the rule rather than the exception. Today we notice that those who are less mature are more selfish than those who are less infantile in their mentality. In nature, it is maturity which is fruitful and generous. In our species maturity also implies a better understanding, which in turn generates compassion and sympathy.

Even today, in a culture dominated by the selfishness of individual, women, and particularly mothers, are more generous and altruistic than men.

That the mind is a major enemy of altruism can be seen by analysing what happened in a Western world dominated by the mind-created concept of Judaeo-

Christianity. Judaeo-Christianity introduced the fear of God and the fear of God's punishment. It keeps us in a state of selfish infantile thinking.

The Judaeo-Christian God also demands a total obedience from His believers which reduces them to an even more infantile state.

I stressed before that it is on account of fear that the efficiency of our limbic system and thus the efficiency of our new brain are reduced or eliminated. Counteracting the efficiency of these two brains, fear reduces or even eradicates the possibility of loving, sharing, nursing, nurturing, taking care of others and helping the young and the old.

Christianity preaches charity but seldom practises it. This is hypocrisy, very damaging to a species for whom education is largely based on imitation.

What is more, Christianity teaches us to love our neighbours in the same way that we love ourselves. Reducing us to a state of infancy, Christianity thus reduces us to selfishness. In infancy we tend to love ourselves selfishly. Loving our neighbours as we love ourselves implies loving them in a selfish way, which means exploiting them. When we are unable to exploit them, we often end up hating them. Litigation between neighbours is the best evidence of the selfish love we cherish for one another.

Christ treats us as if we were obtuse children: he orders us to love our neighbours. He never asked himself if our neighbours would be really happy with our love, with our continuous interference in their lives, with our intrusion into their privacy.

Having been too busy to prove that he was the Son of God, Christ could not have realised that a love for others should be preceded by an *understanding* of others which the self-centred infantile mind is not able to attain. It is only by understanding others that we can develop a true love for them which is in their interest and for which they will be thankful. Their sense of gratitude will make them well-disposed to us and increase their understanding of the way in which we think. It is upon this understanding that their love for us and for their other neighbours might be built.

Profit

As the mind emerged, so did an ever-increasing awareness of the notion of profit.

Being unscrupulous by nature, the mind's pursuit of profit became a veritable thirst for exploitation, a hunger for profiteering from others, a profit addiction.

Any advantage or benefit realised by man would serve to bolster his fragile ego. In fact, the more fragile an ego is, the more it is obsessed with the pursuit of profit.

The pursuit of profit is in the province of the infantile mind, and therefore has no limits: often it grows exponentially.

People who are obsessed with the pursuit of profit consider it as an award or reward for cleverness or deviousness. They tend to accumulate profit even when they no longer need additional wealth. Through accumulating riches they hope to acquire importance and power, the supreme aspirations of the fragile ego. Often, however, an insecure ego realises that even when it has attained a position of importance and power, it is ironically even more fragile and insecure than it was at the outset. It then develops a fear of losing everything.

The mind's obsession with profit can end with the disease of cupidity.

The most ugly and damaging aspect of profiteering is that it is often conducted at the expense of other members of the same community, often at the expense of members of the same family.

The more fragile a man's ego, the more ruthless and competitive he is in the pursuit of profit. The more ruthlessly competitive he is, the better chance he has of becoming successful, of being considered the *victor ludorum*. The more intelligent, humane, understanding, altruistic and compassionate a person is, the less successful he is in the pursuit of profit and wealth. The demi-gods of the western capitalist world therefore hold him in low esteem. I wonder how this fits into the Darwinian theory of natural selection? Does natural selection consist, in fact, of regressing us to our previous reptilian existence?

Fear of Death

One of the greatest illusions or self-deceptions practised by the mind is its belief in eternal life. Many believers spend their entire lives entertaining this fixed illusion. Burdening our psyche with lies creates tension. This tension prevents us improving the quality of our earthly life.

The Christian belief in life after death instils fear of such concepts as the Last Judgement and Eternal Damnation. Equipped with crucifixes and other *memento mori* our religious leaders remind us of our impending death and the possibly tragic eternity of the afterlife. They try to transform us into slaves, into obedient children.

If we did not believe in life after death, we might be more serene, more humble, more humane and more settled in this life. If we did not believe in life after death, we would eliminate religious fanatics who believe they will reach an eternal paradise if they die fighting the so-called infidel. Other religious fanatics believe that they too will reach paradise if they kill the enemies of their religion.

The Roman Stoics insisted that only by not fearing death can we achieve a life of dignity, serenity and maturity. *"Qui mori didicit, servire didicit"* stressed Seneca, explaining that those who do not fear death can live a life of dignity. This is, perhaps, why the first Christian scholars hated the Stoics. In fact, St Augustine gathered Christian scholars around him to write a book which set out to denigrate the virtues of Ancient Rome.

Since he has been able to think and to reason, man has tried to avoid death. When they observed nature our primitive ancestors saw seeds falling in the autumn and reappearing as plants the next spring. Heartened by this regenerative process, they started burying their dead, hoping to see them reappear. When the departed failed to appear, they were forced to invent an underworld similar to the visible world. When they realised that no-one ever came back from this underworld and that bodies merely disintegrated, they were obliged to invent the 'eternal soul', which goes either to Paradise or to Hell, depending on the moral behaviour of the bearer.

Such was the wishful desire for an eternal soul that we never asked ourselves: what is this soul? To which part of the body or the brain is it related? Is this soul able to sense or perceive things? Can this soul have feelings and memory? In fact, since it has no memory, the soul is a pure abstraction. In order to have a memory the soul would have to be connected to a living central nervous system. What is more, if the soul possessed memory, it could not be happy in paradise; memory, confronted by the boring and monotonous prospect of eternity, would incline to develop a nostalgic yearning for a life of changes.

The idea of the soul was created by the primitive reasoning of our ancestors, some thousands of years ago. To believe in its existence today implies regressing our reasoning power to the primitive levels of our ancestors.

Pope John Paul II stresses: "Immortality is not a part of this world. It can come to man exclusively from God." Given the fact that immortality or eternal life are contrary to existence and given the fact that they can be given only by God in Heaven, we can deduce that, being immortal and eternal, God and His Heaven and Hell must not exist. Eternity cannot exist because whatever exists is not eternal. Eternity is supposed to be static and immutable while life and the universe are in a permanent state of becoming, a dynamic state of evolution and change.

When God, through our religious leaders, explained that His obedient children would enjoy eternal bliss in Paradise and that the miscreants would suffer in Hell, He did not take into account that in eternity, if eternity existed, everything would be the same, and that in eternity there is no such thing as blissfulness or suffering, because in eternity there are no changes.

"God so loved the world that He gave His only son, so that everyone who believes in Him might not perish but might have eternal life." These words from St John's Gospel show us that *if eternal life could not exist, then God's sacrificing His own son was an act of futility.*

Intuition and the Mind

We often do things without consciously knowing about them. This unconscious activity is ruled mainly by what we call intuition.

A natural defence mechanism, instincts and intuition are the primordial parts of any organism. They are all inspired and generated by the main tendency of organic matter which consists of trying to reduce its existential instability or vulnerability and to avoid any worsening of its existing uncertainty and frailty.

Each organic molecule seems to carry an innate tendency to return to the less unstable or less precarious state of a previous existence. After all, this should be understandable if we take into consideration that life is imposed by solar energy on inorganic molecules.

These tendencies of organic compounds to reduce their instability must have inspired the formation of the first cells, this basic unit of life. The first part of the cell to be formed must have been its membrane. The cell's membrane must have assumed a round or globular form in order to find a less vulnerable existence, a less unstable state.

Inside these primordial bubbles, a variety of organic molecules must have found a protective shelter. Pushed by the tendency of finding a reduced instability these organic molecules inside the bubbles discovered co-operation. In this biochemical process complementarity played a major role. Any uncooperative organic molecule would have been ejected by the co-operation of molecules inside the cell. In fact, any living cell is nothing else than the organisation of co-operation of the unstable organic molecules in search of reduced instability, reduced irritation, reduced discomfort. Any multi-cellular organism is nothing more than the organised co-operation of cells in the pursuit of a reduced instability, a reduced disorder.

Gangs, groups or societies composed of individuals are inspired by the tendency of individuals to discover, through togetherness, contract and co-operation with the other individuals, a lower state of instability or uncertainty.

Seeking decreased instability, organic compounds tend to create more complex forms different from the sum of the compounds composing them.

Growth in the living world is generated by the instability and vulnerability that the living matter carries inside itself. Only an unstable cell under pressure will divide into two cells. A cell in a stable and comfortable state of existence will never divide.

The great irony of life, however, is that the more complex form of life formed by the restlessness or agitation of cells in search of a reduced instability, is even more unstable than the simple forms of life which compose it. Any increase in complexity increases its needs, its vulnerability and its precariousness. The simplest forms of life, which are bacteria, prospered on our planet billions of years before any multi-cellular organism appeared, and most probably will prosper billions of years after *homo sapiens* and many other multi-cellular organisms disappear. This should not be surprising if we take into consideration that multicellular organisms have been formed by more fragile and more unstable bacteria, bacteria in more urgent need of complementarity.

Instability provides vibrations and vibrations produce waves. Through these waves everything in life interacts with everything else. That is, perhaps, why we can sometimes get the feeling that something of which we are not consciously aware is about to happen. That is, perhaps, also why wishfulness or desire, which are inherent in the instability or discomfort of the living organism, can influence or predispose events in their favour; that is, perhaps, why wishfulness and desires can help our premonition or intuition to become reality.

Intuition can also be inspired by an organism's unconscious perception of events taking place inside its own body and particularly of events taking place in the organism's natural defences which are, after all, intimately related to the intuition as they also are inspired by the main tendency of the living world: reducing existential instability and preventing its increase.

Sooner or later we may invent special instruments which will help to read the waves which emanate mainly by organism's immune and repair systems. Reading these waves, specialists will be able to diagnose hidden diseases and weaknesses or the predisposition to certain physical or mental disorders,

Being brought about by the living organism's instability, uncertainty, fear, tensions and anxiety, its

94

unconscious perception or intuition will increase or decrease with the waxing or waning of the organism's instability, uncertainty, fear, tensions and anxiety. This can best be seen in the case of mothers. Having children, and thus acquiring new fears and worries, mothers become more intuitive.

That intuition is intimately related to the organism's instability and uncertainty can be deduced by considering the fact that unconscious perception is an extra activity which requires more energy, which in the living world can be provided mainly by extra instability or extra fears. There can be no intuition in equilibrium, in stability or in static existence. That intuition is related to instability and uncertainty can also be deduced by the fact that our unconscious attention and perceptiveness can be provided only by instability, uncertainty or frailty.

One of the primordial tendencies of any living organism is to establish its proximity or its distance with the world around it following the rule of reduced instability as in the case of the proximity, or trying to prevent the increase of instability as in the case of distance from anything that can be harmful of damaging. With the appearance of the first organic compounds came attraction and repulsion in the living world. It is a well known fact that bacteria tend to migrate towards conditions guaranteed to reduce their instability or discomfort and to migrate away from conditions that can aggravate their existence; that can be irritating. Could it not be that even the antigen in an organism is attracted by the organism's antibodies as antibodies can reduce the antigen's instability, potency and virulence?

We are intuitively attracted by beauty, harmony, grace, cooperativeness, peace, generosity, love or kindness because they reduce our existential instability and fears, they are not threatening. On the other side, we try to avoid ugliness, freaks of nature, violence, aggression, chaos or disorder for the opposite reasons.

It is our intuition which is intimately related to our immune and repair systems and their potential and efficiency which inclines towards influencing our lives, our choice of partners, of our friends, of our profession, of our hobbies, of our aspirations, and of our taste. It is our particularly frail species, in need of protective intimacy, togetherness and belonging, which was able to develop our intuitive pity, empathy, sympathy, compassion and love.

Being inherent in organic compounds, intuitive potential must play an important role in the formation of an organism's senses, its perception and the brain's reasoning. In fact when we have difficulties in our conscious problem-solving we lean on our intuitive intelligence.

Intuitive intelligence can be deeper and more universal than the reasoning of our brain. This is perhaps why many universally valid discoveries were divined by intuition. The theory of Quantum Mechanics can well illustrate these thoughts.

Since it is intuition which influences our attitudes and behaviour to a great extent, our pretentious belief in individual free will, of which we are so proud, seems to be an illusion, perhaps even a delusion.

Intuition, this innate intelligence, which has assisted our species for millions of years of life in the unpredictable and unfriendly savannah, started being confused and frustrated by the development of the mind. With its fixed ideas, its blind faith, its rigid stubbornness, its strong preconceptions and prejudices, its fanatical religious or ideological dogmas and its institutionalised values, the mind is able to inhibit and to eliminate intuitive intelligence. In fact being less attracted by the mind's surreal world, woman is more intuitive than man.

One can observe, also, that people with a sense of humour, which derides the mind's world, are more intuitive than those lacking it.

Sleep, Dreams and the Mind

We spend nearly one third of our lives in sleep and we still do not know the real meaning of it. Also, a part of sleep is spent in dreaming which is another enigma.

The following theory might throw a new light on these problems.

As I stressed before, life is not a gift of God, as God, who is supposed to be a loving God, could not have created a suffering world. Life, in essence, consists of painful effort by the unstable organic molecules to search for less instability, less irritation, less discomfort. Implying suffering, life must have been forced on the living world. Life is a forced labour imposed by solar energy. In fact, when man invented the mind and its wishful creativity, he invented life after death, life in paradise, life without suffering. In reality, life is a sad phenomenon. Sensing that life is bound up with suffering, many people complicate and dranatise their existence in order to enrich their lives.

When the sun, which provides the main energy of life, is less powerful, less present or when the sun is on the far side of the globe, as during the night, life tends to reduce its restlessness, agitation or liveliness, to generate the least possible activity, to reach a dormant state. In fact, most animals and plants sleep during the night. Appearing again the day after, the catalytic sunlight interrupts sleep and revives life. It is interesting to notice that when an animal senses the approach of death, it tends to hide in some dark or shady place in order to avoid further irritation, to die more peacefully.

Sleep is the nearest state to the stability of inorganic matter that an organism can reach. Because it is imposed by external forces, life cannot annihilate itself by itself, but in the case of man, and only after he had discovered his mind: man is the only animal practising suicide.

Influencing our brain's development, its organisation and its activities, our intuition and our natural defence systems can also affect our sleep and our dreams. Informing the brain of their extra activity in the organism, the defence systems tend to stimulate the brain's sleep mechanism. When in its developing phase or in its early infancy, with the defence systems more active, an organism tends to spend most of its time sleeping. In adulthood, and particularly in old age, when the immune and repair systems are less efficient and less active, sleep is noticeably reduced.

Most people consider tiredness as the major cause of sleep and sleepiness. Tiredness, in fact, revives our defence systems, which, through their increased activities, stimulate sleep and sleepiness.

As I stressed before, man's mind and his defence systems are related to each other. The mind can improve the efficiency of our immune and repair systems. This can best be seen in the beneficial effect of placebo drugs on our organism's natural defences.

There is evidence also that a frustrated mind, a mind in a state of emergency, a mind threatened in its strong beliefs, can inhibit the efficiency of our natural defences. Over-active immune and repair systems can limit and even eliminate the mind's aggression, adventurousness, competitive spirit, erotic imagery, pursuit of excitements and man's ego's infatuation. A certain range of the lasting activities of the defence systems can give rise to chronic depression or melancholic mentality.

As, during sleep our intuition and our defence systems are active, they can influence our dreams.

That dreams can be under the influence of our intuition and of our natural defences can be deduced from the fact that human and animal young, whose intuition and natural defence systems are particularly active, spend a great deal of their sleeping periods, dreaming.

That dreams can be influenced by our immune and repair systems can be deduced from the fact that sometimes, if correctly interpreted, dreams can tell us about our organism's conditions and health, they can inform us of an existing latent or silent disease, and they can even foretell the dreamer's death.

The origin and the nature of dreams could be explained with the following theory.

Our memory which plays the essential part in our dreams is a result of an activity. An activity is accompanied by an energy. The energy which helps creating memory must be provided by a sensation or an emotional arousal, formed by real experience or by the mind's fantasies and imaginings. The real experience or the mind's fantasies and imaginings are registered in our brain's memory-pool by the frequency or intensity of sensation or emotional arousal that they create. The stronger the sensations or emotions are, the stronger and more deeply-ingrained the memory of events that provoked them will be. Because young people experience sensations and emotions more intensely, they are better at memorising persons or events than older people.

Any new sensation or nervous arousal will register in the memory-pool the events which cause it on the intensity or frequency of energy it carries. These new experiences will unite with the variety of experiences already registered in the memory-pool on the same intensity or the same frequency of the energy on which *they* had previously been registered. There are living entities we encounter at regular intervals: a friend, a dog, a shop assistant or whatever. In each case the encounter is registered in terms of the emotional arousal or stimulus it provokes.

Our intuition and our body's defence systems can also provide sensations and emotional arousal. When, in the course of sleep, a new sensation or a new emotional arousal, caused by the intuition or defence systems, presents itself in our memory pool, it revives a chain of images of our past real experiences and of the past experiences of the fantasy world of our mind already registered on that same frequency in our memory. Each sensation or emotion created by our intuition or by our defence systems carries its own frequency, its own intensity.

In the absence of the rationale and logical structure of our conscious brain, these revived images in our memory by the sensations or emotional arousal caused by the intuition or the defence systems, give the impression of being incoherent, incongruous and irrational. However irrational and incoherent our dreams might appear to be, it is clear that they have their own internal logic, their own congruity which, I suppose, can be found in the organism's tendency to diminish its instability.

A mother's intuition and the activity of her natural defences must play an essential part in her foetus's dreams. A mother's intuition and the activities of her immune and repair systems must be reflected in the form of the images in the brain of her foetus.

The influence of the body's internal activities in dreaming can best be seen in the increase in frequency and intensity of a woman's dreams during pregnancy when her organism is facing important hormonal changes.

When during sleep our intuition becomes particularly active or when our defence systems face some special difficulty or problem, we may have nightmares.

Most probably these extra difficulties in our defence systems, or of our negative intuition, stimulate the mechanism in our brain dealing with states of emergency. In fact, during most nightmares we try to fight, to flee or to hide, all of which are intuitive reactions to a state of emergency or fear.

These nightmares can range from panic and paralysis to hallucinations and out-of-body experiences, from semi-consciousness to altered states of consciousness, from a confusion of dreams with day-dreams to straightforward consternation.

That nightmares can be related to our intuition and innate defences can be deduced by the fact that they exist in all cultures.

In most nightmares we are threatened, terrorised, attacked or in dread of being attacked by extraordinary, mysterious, surreal people or events. Strong emotional arousal created by the emergency mechanism, provoked by the extra activity of our defence systems or a negative intuition, can revive in our memory pool the network in which terrifying, mysterious and surreal people or events were previously registered. Many of these chimera are registered in our memory pool by sensations or emotions created by our mind's fantasies or imaginings in our infancy in which threatening and terrifying supernatural people and events, demons and monsters played an important part.

Many people have nightmares when they are run down or in a frail state. It could be anything from indigestion or food-poisoning to excess of alcohol or narcotic drugs.

Humour and the Mind

With the appearance of the mind appeared ridicule, which illustrates the rear nature of the mind. The mind's world is not a serious world, it may take itself seriously, it believes itself to be serious, which thus makes it over-serious and that over-seriousness is not really serious. This is what makes the mind's world laughable.

Man proudly insists that he is the only animal which is able to laugh. Man, however, seldom notices that he is the only laughable animal. (Other animals make us laugh when they remind us of our own ridiculous state).

Our ridicule is mainly caused by irrationalities, incongruities, fantasies and pretensions. These are all the mind's creations. In fact, living more in the world of the mind than woman, man is more ridiculous and comic than woman. This is, perhaps, why we have more male than female clowns and comedians.

It is our mind's system of beliefs which is ridiculous. Building castles in the clouds and living in them is laughable. In fact, the best therapy against many mental disorders is humour.

Religions and ideologies are an important source of ridicule. Religious and ideological believers, in fact, fear a sense of humour or laughter because they deride their beliefs. This is why laughter and humour are often persecuted by despotic religious and ideological regimes.

When the philosophy in Ancient Greece was at its zenith, some of its adherents considered laughter as a sign of madness, such was their fear of being derided.

Medieval Christians considered laughter an affront to the deadening over-seriousness of their assumed religiosity.

The absurdity of Christian beliefs can be deduced from the following plethora of incongruities.

We are indoctrinated into believing that our God is a loving God. On reading the Bible we discover, instead, that He is desperate for our love and our worship. This would imply that God is insecure and dependent on our love and worship. In insecurity and dependence, God and man, (who is made in God's image) are unable to love anyone but themselves.

The Bible also teaches us that our God is an omnipotent God. But if our God was omnipotent, He surely would not need our love and worship.

We are told by the Bible that this omnipotent God created the World. Christian believers do not realise that if anyone ever achieved omnipotence he or she would not be creative. Omnipotence must be sterile as it has no needs or desires. In fact, the idea of omnipotence could only have been invented by an infantile, undeveloped or impotent mind.

We are also told that our God is omniscient. We are informed by the Bible, however, that 'The Lord, thy God, is a jealous God.' A jealous God cannot be omniscient because jealousy limits the efficiency of the brain and its memory.

We read in the Bible that 'He that increases knowledge increases sorrow.' This must imply that our omniscient God is very sad indeed. It also seems incongruous that we carry on confessing our sins to a God who, in His omniscience, should already be aware of them.

Christian leaders never cease their litany that Christ saved us from Original Sin. God cannot punish curiosity and a desire to explore, which are the main characteristics of our species, particularly in our infancy; and Adam and Eve were infants when God created them. As God created us with these characteristics, God cannot punish us for exercising them.

Christian leaders preach that we should believe in Original Sin and believe that a "merciful" God sent His only begotten son to save us from this "terrible sin". But what about the majority of humanity who do not believe in Christ and who have never heard of Original Sin? It seems that with their Christian God, Christians occupy a privileged position, since after their death they can look down from their Heaven on all those "infidels" or non-Christians burning in hell. What Heaven!

The story of Original Sin is an infantile joke. First God tell us that we have committed this Original Sin, then he sends His son to save us from the same sin. Would it not be more logical if, instead of sending His son to be crucified in order to save us from an invented sin, He had sent another, more cheerful son to announce to the world that Original Sin had never existed, that it is in the nature of infancy and an infantile mentality to be curious and exploratory, to learn things, to eat the fruit of knowledge, as learning can only improve the quality of life?

Infantile curiosity and exploration are innocent games. Condemning and punishing such games can only be done by a father who is afraid to allow his children to free themselves from a father's influence.

If God had sent a cheerful prophet, or a second son to justify the exploration of the Tree of Knowledge, we would have had, and we would now have, a happier life: and in this happier life we would be more generous, more humane, more caring and more loving. Without Christ's crucifixion we would be far better Christians.

In legal and moral terms we should not be punished for an inherited sin as sin cannot be inherited. To accuse someone of a sin or a crime which he or she did not commit is a serious slander.

There is something particularly humorous about Original Sin. If we are born with Original Sin, if we come into this world as genetic sinners, then it seems that it would perhaps have pleased God if we had not been born at all. It should be understood that God does not like sinners. If God does not like sin and sinners then it must be the greatest sin of all to procreate sinners. If that is so, then the Pope should bless abortion and contraception as they prevent a swelling of the sinful ranks. But perhaps God likes sinners, for without sin and sinners there would be no need for Him. If this is true then the greatest sinners are Catholic priests and nuns, as they are supposed not to beget sinners.

But the most humorous aspect of Original Sin is the fact that for millennia woman has been considered guilty and responsible for the loss of Paradise. If we read the Bible properly, we will discover that Eve was created by God after God told Adam not to touch the Tree of Knowledge. God never told Eve not to eat the fruit from the Forbidden Tree. Most probably, excited by the creation of Eve, and longing to play with her, Adam forgot to warn her about the Tree of Knowledge. After all, Adam had only recently been created, and memory is short in infancy.

Women should be proud of Eve and her faculties of common sense and logic. She must have sensed that it was contrary to nature to have a tree with such nourishing fruit and not to eat it. If God had not wanted Adam and Eve to commit Original Sin, He would have created a fruitless tree or would not have created a Forbidden Tree at all. In fact, Adam and Eve are not guilty for having eaten the forbidden fruit: God is guilty for having created such a tree. Our fearful God must have created the Forbidden Tree in order to please His enemy, Satan.

In relation to Original Sin, we discover another silly incongruity. Why did our "just" and "merciful" God have to punish other animals, living in peace and harmony, by throwing them out of His Paradise, and transforming them into wild beasts, simply because of Adam's and Eve's Original Sin?

We have been taught that God created man out of love.

Could it not be that God created man in order to discover Himself, to discover who He was, what He looked like? That is perhaps the reason He created man "in His own image and likeness". God has no way of discovering who He is or what He looks like as He is alone in the universe. He had no reflection, no mirror which could have helped Him to discover Himself.

But could God have created man in His own image when He had no idea who He Himself was?

There is another enigma: if God created man in order to discover His own image, why did He then have to create woman? It seems that God created woman in order to provide Adam with company. This might lead us to the conclusion that the primordial God's feeling was one of loneliness. Only a feeling of loneliness could have implanted in God's mind the idea of company. It was this need for company which made both Man and God dependent beings: God becoming dependent on man and man on woman. Woman's loving, nurturing, caring and mothering place her in a position of superiority.

God could not have created man out of love because He was alone in the Universe. The idea of love and loving has to be inspired or suggested by a woman, by a mother. God has no mother.

Throughout the Bible God speaks to His people.

In which language did God address His people?

We know that language is intimately related to the community and the need to communicate among members of that community. There is no language which can be learned in solitude. Only with the appearance of Eve could Adam and God have learned how to communicate, how to speak. If God was able to speak, then His mother language must have been that of Eve.

But could God have learned how to speak at all?

In order to learn a language one has to listen. An omnipotent and omniscient God is unable to listen. He cannot even hear as He is self-absorbed. Even if God created the world out of love, there is evidence that He did not like what He created. We read in the Bible: "And God said: "I will blot out from the face of the earth all mankind that I created. Yes, and the animals too, and the reptiles and the birds. For I am sorry I made them."

If God did not like the man He created in His own image, it might well mean that He did not like Himself. Perhaps He did not like Himself because, like any infant, He felt incomplete, immature, in need of love.

Most people know about Christ's good deeds, especially his curing of invalids and most notably in this respect his spectacular revival of Lazarus. We never ask what happened to these people after Jesus' miraculous performances. We might profitably dwell on the notion that after having been resuscitated by Jesus, Lazarus might have led a thoroughly miserable life. To have been old, poor and decrepit in those days, in an occupied Palestine, must not have been an easy life. There was no pension or free medicine or Medicare in those days. Even if these services had existed, Lazarus would not have been entitled to them any more because he was officially dead. What is more, Lazarus would have had to face death for a second time.

But what is really comic is the fact that if Lazarus was such a good person as to deserve resurrection, then he must have been in Heaven enjoying his after-life. If this was the case, Jesus hardly did Lazarus a great service by wrenching him back from his happy life in Heaven to a life of hell on earth.

The main Christian miracle, that of the "Virgin Birth", is particularly absurd.

If Christ was born of a virgin, then he would have been a female: he could not have had male chromosomes, as the male chromosomes can only be transmitted by the father's sperm. If, on the other hand, Christ's mother, Mary, conceived the child with the "Holy Spirit", as many believe, then again Christ must have been a female as the Hebrew term for Spirit, which is *"Ruah"* is feminine.

What Christian leaders never realised is the fact that through his miracles Jesus committed a serious sin against his Father. In performing his miracles, which are contrary to the natural laws and order created by our God, Jesus was arrogantly placing himself above the supreme Creator. In a certain sense, Christ was playing practical jokes on his Father's world and its laws. I am sure that God, who has never revealed any sense of humour, was not amused by His Son's bemusing miracles. But like any father, particularly as the father of an only son, our Lord tolerated or pretended to ignore, or was even proud of His Son's practical jokes.

During a wedding in the village of Cana in Galilee where Jesus was invited as a guest, the wine ran out. Jesus performed a miracle, transforming water into wine. As he was the Son of an omniscient Father, Jesus should have known that there is nothing more violent and degrading than the behaviour of poor people when they get drunk.

Jesus must have known that drinking is not a wise thing to do and that to provide an abundant quantity of free wine was even less intelligent, because he could have read in the Old Testament that "Wine is a mocker, strong wine is raging".

The evangelist John insists in his Gospel that it was Mary. Jesus's mother, who asked her son to transform water into wine. St John did not know that no woman would have asked such a thing. Women know that they are first on the drunken man's hit list. Drunks, particularly primitive drunks, derive special pleasure from beating their wives, often accusing them of being the cause of our having lost God's paradise, in which wine was abundant. It would have been very sad if the newly-wedded spouse of Cana had ended up battered on the first day of married life.

I am sure that, as a wise Jewish mother, Mary would have asked her son to transform the existing wine into water as she knew that, with plenty of wine around, the wedding would have ended in ugly confusion. As a wise mother, Mary must have known that for a woman her wedding day is the most serious and important day of her life, which should not be transformed into violence and brutality by a lot of drunken men.

One of the most absurd ideas invented by the human mind must be that of Abraham's readiness to sacrifice his son in order to please his God. Inspired by this idea, which is glorified in the Bible, many sons have been sacrificed by their fathers in order to satisfy their ideological or religious beliefs.

Ideologies can be as ridiculous as religious beliefs. We have seen how pathetic and ridiculous those who believed in the Communism have been. More and more we are realising how pathetic and ridiculous are those who still believe in the Western concept of capitalism.

Ideally, we could have witnessed the pathetic and comic side of another ideology, that of the feminist belief in the equality of sexes. Blinded by ideology, many women did not realise how absurd it was to try to be equal to men in a man's world, a world made by men in their own image and in their own interest. Blinded by their ideology, women did not realise how ridiculous they became on attaining equal playing power in the games created by men for men.

But, what it seems to be the most ridiculous is the following fact. In a world created by men, men have a privileged position attained at the expense of women. Women's desire to become equal in a man's world implied that they also desire to attain a privileged position at the expense of their sisters!

Women should not ask for sexual equality in a world created by men for men: women should try to create a new culture, a new world based on their superior attributes and more rational values, and then invite men to be equal to them, if they are capable of it.

Feminists are mostly adolescent in mentality. There is no difference between the male and the female adolescent mentality: both of them are characterised by a blind faith in wishful beliefs of superiority which give rise to arrogance, aggression, self-righteousness, fanaticism and militancy. This, perhaps, is what impels the adolescent-minded female to strive for equality with men. The fundamental difference between men and women in the context of the adolescent mentality is that while women are able, through maternity, to evolve from the adolescent mentality to maturity, men tend to remain stuck for the rest of their lives. In all other animal life there is no such conundrum, since the adolescent stage is transitory, straightforward and unencumbered by emotional complications.

The most incongruous side of the mind, however, seems to reveal itself in the fact that it has no interest in helping our brain to reach serenity, intelligence and wisdom as these limit the freedom of the mind's fantasising which is exciting and amusing. The infantile mind finds serenity, intelligence and wisdom boring.

Other Consequences of the Mind

With the appearance of the mind, we added the mind's suffering to physical pain. Physical pain tends to offend or to irritate the mind's ego and this generates suffering which can aggravate pain. In fact, having a less inflated ego than man, woman copes better than man with physical pain. Suffering can also prevent or slow down recovery from the causes of pain.

The mind also often develops self-pity which can damage our physical and mental health.

Being a creation of the mind, suffering and self-pity are closely related to culture and to religious or other beliefs.

The mind introduced envy, jealousy, anger and bitterness, all of which are unhealthy.

In order to protect its frailty the mind invented pride which easily inspires the mind's ugly and aggressive scorn for whoever or whatever threatens its mind's *amour propre*.

With the development of the mind human language started growing proportionally. The mind introduced a new range of emotions. The mind's craving to materialise in some way its imagination and its fantasies led to the sophisticated development of language.

Many insist that by enriching the mind, language improved our communication. But, we also have to take into consideration the fact that, belonging mainly to an infantile mentality, the mind also uses the language for disinformation, lies, verbal seduction, deception, cheating, and pretence. These characteristics help the mind to create what it considers its supreme virtues, which are its cleverness and its deviousness.

The development of the mind might have contributed to the increase of Alzheimer's disease. The more we are curious and the more we exercise our mental activities, the more efficient our brain is.

With the mind's religious and ideological beliefs, its fanatical ideas, its prejudices and its dogmas, the brain's curiosity and utilisation are reduced which in turn accelerates atrophy, thus speeding up senility.

Expansion and growth, these attributes of infancy, became the mind's obsessions. The mind became attracted to grandiosity. 'Think Big' became the supreme value.

The mind's obsession with growth can best be seen in the uncontrolled pursuit of economic wealth. Like any other expansion, economic growth increases instability and precariousness. Economic growth also increases people's discontent as it increases their pretentiousness and envy. This is very evident in the economically developed countries in the last fifty years. Statistics show that from 1950 to the end of the 20th Century economic prosperity and the material standard of living rose in these countries in a remarkable way. In the same period, in the same countries, people's unhappiness and depression, psychosomatic diseases and disorders increased in proportion. In the same period of the economic growth these countries realised a parallel growth in drug abuse, alcoholism, crime, violence, suicide, eating disorders, obesity, adventurousness and gambling. These are all signs of people's instability or uncertainty, discontent or dissatisfaction.

If our rational brain were to liberate itself from the mind, it would also be able to realise that excessive economic growth, coupled with the uncontrolled expansion of population, is exhausting the economic potential of our planet placing at risk the very survival of life itself.

Our mind's wishful beliefs are often so strong as to be able to influence even science. The monotheistic idea that there is a one and only God who created everything and everybody might have given rise to reductionism in science, reductionism implying the tendency to explain complexities with a single part of the whole. Presently we see biology taken over by the myth of the omnipotent, omnipresent and the one and only dominant gene. Those who explain everything with the selfish and self-centred gene seem to avoid noticing that genes tend to be in a dormant state and that their activity has to be stimulated by the cell which is under the influence of its environment.

There is positive evidence that starvation, fears, anxieties, insecurity, precariousness, diet, drugs, alcohol, hormones, vitamins, climate, environment and the epigenetic instruction can strongly influence performances and activities of our genes.

Science, which is still dominated by men and the male mentality, tends to hold that human conception takes place when man's "fittest" sperm succeeds "in penetrating" the female egg. Would it not be more accurate to conjecture that it is in fact the egg which chooses the most complementary sperm from amongst the many thousands which assail it? This does not imply that the sperm absorbed by the egg is the fittest, the healthiest or the most aggressive of all those sperm which succeed in reaching the egg. If the sperm "penetrated" the egg without the egg absorbing it, the egg would reject it as a foreign body. After all, this would be more in tune with the basic physical and biological laws as the female egg is far larger and stronger than the man's sperm.

Marx explained God's omnipotence in terms of economics, Freud in terms of sexuality.

In its quest for its own importance, our speculative mind stressed and continues to stress that the main purpose of male life is to ensure the continuation of his genetic makeup and its subsequent embodiment in new generations.

Created and conditioned by solar energy and the specific climate of our planet, life cannot have its own purpose, its own will: it has to obey pre-imposed rules.

In specific seasonal conditions sex glands release virulent sex cells which create irritation. We call that irritation sexual arousal. During this sexual arousal our organism is desperate to placate or eliminate the irritation. This is achieved primarily through sexual intercourse. Sexual intercourse serves to eliminate the irritation caused by the sex cells. In fact, the extra energy needed for sexual activity is provided by the biological discomfort created by the virulence of the body's sex cells. Procreation is not the purpose of copulation but, rather, an *incidental* aspect of sexual intercourse. We frequently use contraceptives to prevent pregnancy and the subsequent perpetuation of our genes. Most males in nature tend to abandon their females when they have eliminated the irritation brought about by sexual arousal. Most males in nature take no part in the nurture of offspring. This all serves to render absurd the notion that the primary purpose of sexual intercourse is to perpetuate the genes.

Non-reproductive homosexual behaviour in many species also gives the lie to the theory that the sole purpose of copulation is procreation.

Creating its own idealised world, the mind developed a hostile attitude towards the real world of nature. Only in this way can we understand man's systematic and often enjoyed, pollution and destruction of his natural environment. We consider nature as if it was not a part of us. We even fear nature as it is threatening our mind's world.

Only the mind's hostility towards the real world of nature can explain its obsessive pursuit of the genetic manipulation and cloning in spite of the scientific evidence that genetic diversity can be an advantage.

It is, perhaps, this hostility of the mind toward the natural world which has contributed to the development of the abstract arts which try to deform, to debase and to destroy the real world.

Another profoundly destructive consequence of the mind at work is its ability to create what might be termed 'mental fever'. The mind is able to produce a feverish state of excitement and agitation which can create physiological reactions similar to symptoms of real fever, such as a quickening pulse rate, higher blood pressure, hot and dry skin, poor digestion, impaired efficiency of the senses, perception and rational reasoning, hallucinations and ecstasy.

Our mind can seldom create a peaceful and respectful coexistence with our fellow man because, in common with our God, we tend to force others to fit in with our own wishful image of ourselves.

With the development of our mind, which was guided principally by our inadequacies and shortcomings, we developed another uniquely human trait, the feeling of shame, shame at our immaturity. This impels us to try to conceal what we consider to be our underdevelopment. We are ashamed to be seen naked in public, to be seen with our sexual organs uncovered, to make love in front of other people. We consider our passions to be immature and we therefore make every effort to hide them from view. The development of the mind led to our awareness of our immaturity, and this in turn led to a loss of innocence, a loss of spontaneity.

The Best Way to Achieve Maturity

Science is discovering every day more and more that the female brain is better organised, more integrated into nature and its order and more mature than that of man. In the light of this scientific evidence, why, after so many centuries of disastrous experiences caused by man's mind and his brain, can we not start using the qualities of woman's brain and woman's innate maternal attributes such as loving, caring, nurturing, nursing; sense of organisation, co-operation, patience, tolerance, sense of responsibility, her aversion to pollution and her sense of future?

A woman also has a better capacity for understanding which is essential to reach the sympathy, empathy, pity and compassion so much needed given the continuous increase in the number of the old people.

In order to improve our future and to create a better quality of life, we should place these maternal qualities, many of which were debased, banalised or derided for millennia by male Judaeo-Christianity, into the supreme values of the new maternalised culture. The only way for puerile-minded man to improve his and his planet's future is to grow up culturally, to become mature by becoming maternal.

In order to help us make the first step towards maturity, our new culture should explain that the present male culture's values, such as selfishness, self-centredness, self-interest, competition, aggression, and adventurousness belong to the reptilian legacy which is prominent in infancy, a part of the infantile mentality.

If the male hormones which make people selfish, aggressive and competitive can be increased and kept at their high levels by man's mind's wishful beliefs and fantasies, then these male hormones can also be reduced to lower levels by the brain's rational capacity or by adopting any of the aforementioned maternal qualities. In fact, a loving or a caring man can very well become less reptilian in his mentality.

If the level of male hormones can be increased by an offended ego or by a frustrated mind, then the level of these hormones can easily be reduced to lower levels by our sense of humour or of self-irony which derides our inflated ego or our pretentious mind.

Science informs us that our brain is already organised by male and female hormones when we are in the womb. This implies that mother's hormones must influence the process of the formation and the organisation of the foetal brain. This seems to imply also that a mother's male and female hormones can influence in the foetus the degree of its masculinity or of its femininity. It seems, therefore, logical to suppose, that, if during pregnancy a mother lives in a state of stress, anxiety or fear, she would have in her body more male hormones caused by emergencies. It seems logical also that with a higher level of male aggressive hormones in her body, a mother will transmit a higher quantity of these hormones to her child's brain making it more predisposed for the rest of its life to fears, stress, stress-related disorders, high blood-pressure, selfishness, aggression and violence.

If all this is plausible, then, by maternalising our culture, by transforming the mother's attributes into virtues, by debasing or by deriding man's infantile aggression, violence and selfishness, by playing down the cult of youth and the pursuit of excitement, we might create a more salubrious and more civilised cultural climate: we might have the makings of a new generation that is less aggressive, less violent, more humane, more tolerant, more co-operative and more rational. One notes that many people whose mother's pregnancies took place during a war, a revolution or when in a concentration-camp, tend to be more fragile, more selfish and more aggressive than those whose pregnancies took place in more harmonious times and circumstances. In fact, communities will gain a great deal if women condemned to prison have their sentences suspended during their pregnancies.

As many mothers are more anxious during their first pregnancies because of this new experience, first-born children tend to be more aggressive and more self-assertive than their younger brothers or sisters.

Mothers living in precarious or degrading economic or moral conditions tend to produce more aggressive and more violent and cruel children than those living in economically and morally better environments.

Mothers living in areas bordering hostile countries can also produce more aggressive and more violent offspring than those living in safer areas.

An atmosphere impregnated with fear, tension or hatred by ethnic or religious differences and intolerance between neighbouring populations of the former Yugoslavia, for example, must have contributed to the procreation of generation after generation of aggressive and violent fanatics.

An atmosphere of tension created in the areas around the nuclear power-stations or around nuclear reprocessing plants might also increase aggression and stress of the offspring of the sensitive mothers living in these areas.

In the USA, pregnant mothers have to face a tense climate, created by the cult of aggression and by a strong self-assertive way of life. This might explain American children's tendency to become restless, agitated and aggressive.

Many men who are obsessed with sex might fear that the maternalisation of our culture might reduce their sexuality, their macho potency. The new culture will explain that any obsession reduces the efficiency of our senses, of our perception and of our brain's reasoning, limiting in this way our participation in a wider and richer life which can provide far more lasting pleasure than instant orgasms.

It would help the maternalisation of our culture if the inheritance of wealth was transmitted through the female line. This would help mothers live lives less fraught by fear and anxiety and to produce a better human race as a result.

It would also help the maternalisation of our culture if children carried their mothers' names. After all, during conception and pregnancy mothers invest in the new beings far more than fathers. Carrying its mother's name, a child would be nearer to his or her mother and to her natural attributes and values.

Education, particularly that in early infancy, should be in the hands of women and men with maternal qualities. They would thus accompany their teaching with behaviour based on their qualities; they would therefore serve as an example to be imitated.

Teaching by preaching or pontificating, as practised in our present culture, does not work with our imitative species.

In the new culture, teachers should be the best paid professionals in the community. There must be something wrong with the present culture and values if boxers, footballers and in particular financiers (who have occasioned more poverty than wealth in the world) can earn ten or more times than those dedicating their lives and considerable talents to creating a better and more humane succeeding generation.

Three important courses should be introduced in all schools. These courses are: gardening, sailing and a sense of humour. Through gardening and sailing children could learn that in order to advance or to reach some positive result they would have to respect nature and its forces. These two courses could also teach patience, a cure for one of our most damaging derangement, the temptation always to rush: *festina lente*. Many people do not realise that haste and fear are intimately related; in fact, it is fear which provides the energy that fuels undue haste.

With patience we discover time, and with time we discover the future. Discovering the future might impel us to abandon primitive here-and-now gratification and start discovering planning, the lasting pleasure of looking forward. This is particularly important in a world in which contemplative looking forward has been superseded in favour of rushing.

Acquiring a sense of humour could help us liberate ourselves from the incongruities of the mind and its fantasies.

In a maternalised culture children would not be nurtured with fairy-tales, cartoons or virtual-reality. These teach children colourful lies which belong to the mind's world and which increase the mind's predisposition to become attracted and seduced by deception. I am sure that if we were not influenced by fairy-tales in our childhood we would never have accepted ideologies such as communism. Fairy-tales consolidate credulity which in turn perpetuates infancy and the infantile mentality.

If the male state of infancy can be culturally perpetuated by man's male culture then his maturity can be reached and perpetuated by a more mature culture based on maternal values.

Many scientists insist that culture cannot improve genetically selfish man. But, if we take into consideration that it is mainly hormones, instigated by our brain, which activate genes, then we can suppose that while man is in his selfish infantile mentality, it is his male selfish hormones which revive his selfish genes. Perhaps, reaching cultural maturity, man can produce a group of selfless and generous hormones which could revive the activity of selfless and generous genes which are at present in a dormant state? After all, we all carry thousands of dormant genes, some of which could be very generous indeed. These generous genes can easily be seen in many people's altruistic behaviour.

Scientists explain that when she is cuddling her child, a woman's brain and body release certain hormones and neurotransmitters which give her particular pleasure. Scientists also explain that man cannot experience the same pleasure through cuddling a child.

We all know that man gains great pleasure in dealing successfully with things he and his community consider important, things that his culture considers virtues or superior achievements. I am sure that if our new culture were to elevate the care of a child to the status of a superior undertaking, many men would be able to derive great pleasure from cuddling a child rather than the embarrassment they customarily feel. In order to experience pleasure, man has to achieve, to score on the scale of the community's cultural values. He needs a socially recognised success, he loves to be admired, applauded and decorated, he needs endorsement from the majority.

There is more and more evidence that girls' brains are better developed and that they are better in school than boys.

The education programs of the new schools should concentrate on the demasculinisation of boys' brains, on the elimination of the masculine beliefs and prejudices which keep the gates of learning closed. An important aspect of their training, designed to improve their learning capacity, could be in domestic work. Domestic science can help boys, who tend to concentrate on one thing at a time, to widen their curiosity and pay attention to more things at the same time. This would stimulate many parts of the brain simultaneously, enabling it to have a more globally complete vision of reality, to be less unilateral or partial, to be more open to the world and to fantasize less.

Our present male culture glorifies self-confidence. There is, however, an important difference between man's self-confidence and that of a woman. Man's self-confidence is inspired by his mind's pretentious ideas accompanied by aggression, often an irrational or adventurous idea. A woman's self-confidence is based, instead, on her competence. That is, perhaps, why man's self-confidence implies tension, while a woman's is dignified by calmness and serenity. That is, perhaps, why man's self-confidence tends to be intimidating, destructive, offensive and violent while a woman's tends to be relaxing and constructive.

A new, maternalised culture would explain that Darwin's theory of evolution, which glorifies the selfish, ruthless and devious individuals of our species and which considers these individuals to be the fittest of the species seems to be contrary to our interests, to the interest of other creatures and to the detriment of the very life of our planet. To consider individual selfishness, aggression, ruthlessness and deviousness as the supreme qualities is not wise or intelligent because it is these very characteristics which prevent humanity evolving towards maturity, towards a better quality of life, towards a better planet.

In a maternalised culture, maternal attributes such as an aversion to pollution, dirt and filth, would become values to be practised and imitated. Women have always taken the lead as far hygiene is concerned.

Only a culture which cultivates these maternal values could solve the main problem of our planet: pollution; pollution of our water, our soil and our atmosphere.

Women know that an unhealthy and polluted environment in turn produces unhealthy children, children attracted and excited by filth and garbage.

We are entering the Third Millennium guided by a proud and arrogant Western capitalism, proud and arrogant because it has succeeded in imposing itself on the rest of the world.

More and more people are realising, however, that the globalisation of capitalism implies the globalisation of the infantile reptilian attributes consisting mainly of the ruthless exploitation of human and natural resources. More and more people are trying to explain that we need to give capitalism a 'human face'. But, capitalism cannot acquire this 'human face' because, by its very nature, it is reptilian.

With maturity forever beyond its reach, our male infantile reptilian mentality and its burgeoning capitalism will continue to thrive unchecked. Having overgrown itself it will overrun itself. It will decline and collapse, leaving our planet in a desolate state.

Could it not be that the reasoning of this book is futile? After all, it seems that present-day human mentality is fascinated rather than repelled by disaster.

Our infantile mentality is woven from the fabric of fantasies and passions, and these are the enemies of reason.